Polluted Science

The EPA's Campaign to Expand Clean Air Regulations

REVISED AND EXPANDED EDITION

Michael Fumento

The AEI Press

Publisher for the American Enterprise Institute
WASHINGTON, D.C.
1997

12-19-97

Publication of this volume was made possible by a generous grant from the Rollin M. Gerstacker Foundation to the American Enterprise Institute in support of AEI's program of studies in government regulation and legal policy.

Available in the United States from the AEI Press, c/o Publisher Resources Inc., 1224 Heil Quaker Blvd., P.O. Box 7001, La Vergne, TN 37086-7001. Distributed outside the United States by arrangement with Eurospan, 3 Henrietta Street, London WC2E 8LU England.

ISBN 0-8447-4043-8

1 3 5 7 9 10 8 6 4 2

THE AEI PRESS
Publisher for the American Enterprise Institute
1150 17th Street, N.W., Washington, D.C. 20036

Printed in the United States of America

To Billy Richards

Contents

Tables

Figures

Foreword

The Environmental Protection Agency announced in July 1997 that it was tightening national standards for two important measures of air quality—particulate matter and ozone. Hundreds of counties that meet current air quality standards would fail to meet the new ones; as a result, individuals, local governments, and businesses would be obliged to take a variety of steps to reduce air emissions well below current levels. The costs of these measures, in the form of higher prices, lower wages and earnings on investments, and possible "lifestyle changes," would be very high, amounting to billions of dollars per year.

The new EPA standards are not aimed at making observable improvements in air quality—the difference between the proposed standards and current ones can be detected only by sensitive monitoring equipment, and hundreds of such monitors would have to be installed around the country to determine where the standards were being met and where they were not. The ostensible purpose of the standards is rather to promote public health. EPA asserts that they would avert thousands of premature deaths each year and reduce the incidence and severity of asthma and other respiratory ailments. The agency calculates the value of these health benefits to be far greater than the costs of the new standards.

In this critical study of the new EPA regulations, Michael Fumento provides a revealing account of the scientific controversies underlying the particulate and ozone standards and argues that the evidence supporting them is extremely weak and ambiguous. For example, while some studies have found a statistical relationship between elevated levels of airborne particulate matter and increased deaths and hospital admissions, evidence suggests that the relationship is a spurious one—the result of a variety of other factors that would be unaffected by the new standards. Moreover, many other, equally good studies find no relationship at all.

Mr. Fumento criticizes the EPA for mounting a political campaign on behalf of its proposal that grossly exaggerates the likely health benefits and seriously distorts the state of scientific knowledge. The ultimate problem with EPA standard setting, he argues, lies in the Clean Air Act itself, which obliges the agency to search for increasingly arcane evidence of minute health effects from very low levels of air pollution. Yet this is no excuse for arousing unwarranted public fears. The clearest results of adopting the new standards, he concludes, would be to expand EPA's own regulatory powers over states, localities, and the private sector—and to set a precedent for making future revisions to air quality standards with little regard for demonstrated health benefits.

CHRISTOPHER DEMUTH
President, American Enterprise Institute

Preface

The town of Salmon, Idaho, population 3,100, is being done in by two medium-sized fish and one huge federal agency. Ironically, the fish are its very namesakes. The presence of two endangered species of salmon in nearby waters forced the town's only real industry, a sawmill, to close two years ago. Now, says Mayor Stanley Davis, "We're the third endangered salmon." It gets worse. Current Environmental Protection Agency (EPA) air pollution regulations have forced the now-impoverished town to convert one of the school's sawdust-burning heaters to propane, at a cost of a quarter million dollars. The other heater will cost almost half a million dollars to convert—if the town can find the money.

But the worst is yet to come—from new EPA regulations under the Clean Air Act of 1970. The state of Idaho has informed the town that despite its costly remediations, it will not be in compliance with the proposed EPA air pollution regulations. This could potentially bring the wrath of the federal government down on the townsfolk and perhaps force on them such drastic measures as discarding all wood-burning heaters. Salmon's population density, incidentally, is close to zero—one person per 500 acres. Its air pollution problem comes not from smokestacks or exhaust pipes but rather from dirt roads and the occasional forest fire on surrounding federal lands. "It's usually about as pristine as you can get," says Davis. "It's as if the EPA regulations are designed for us not to be here. We just can't comply."[1]

Still, if misery loves company, there will be solace for Salmon. A huge portion of the nation will not be able to comply with what *Industry Week* has called the most "explosive" and perhaps most costly regulations the EPA has ever promulgated."[2] Places that do comply will be burdened with tremendous expense. And a lot of people, including many scientists, are only slightly less outraged or befuddled than Stanley Davis and his neighbors. Their dismay is reinforced by growing doubts about the presumed scientific evidence that underlies and supports the EPA proposal.

Call it the November Surprise. Three weeks after President Clinton secured his second term, the EPA unveiled proposals for sweeping new standards that will once again ratchet down the amount of air pollution allowed in our cities.[3] The agency and its administrator, Carol Browner, claimed the new restrictions would prevent 20,000 premature deaths annually, will head off 500,000 respiratory illnesses, and will improve air quality for 130 million Americans.[4] And all this will cost us essentially nothing. In fact, it will boost the economy! In July 1997, the proposal became a final regulation, barring interference from Congress or lawsuits.[5]

As the old aphorism warns, things that sound too good to be true generally are. What the EPA does not say is that its figures actually represent the top end of a range. And the bottom end? Zero. Zero deaths delayed. Its calculations at this point represent not people but statistical artifacts. Further, they will probably never be anything more. For in actuality, the new standards are based more on whim than on science. The real benefits will be slight at best; more probably, they will make more of us ill and even kill some of us off. They will cost the nation a fortune. And not incidentally, they will extend the EPA's bureaucratic tentacles into new areas that a few years ago would have been unthinkable. One of the nation's leading air pollution experts has called the EPA's decision tantamount to "witchcraft." Yet beyond the issues of science and costs, the debate over the EPA's newest standards raises a much larger question: What does a regulatory bureaucracy do when it starts to run out of new things to regulate? And what do the activist groups that support that bureaucracy do once they have accomplished their stated goals? The answer, it appears, is somehow to fabricate new hazards that nobody previously knew existed, including the regulators and the activist groups themselves.

Acknowledgments

I wish to thank AEI President Chris DeMuth for having taken such a great personal interest in not only pushing but improving this project. In addition to offering valuable ideas, he practically crossed my t's and dotted my i's. Dana Lane, Cheryl Weissman, and the rest of the editing and production department did an excellent job in what I found to be a stunningly short time, as was required by the timeliness of the project. Ed Dale went over the entire manuscript with the proverbial fine-toothed comb and offered many suggestions. Mary Oliver helped with research and she, Matt Kaufman, Charles Freund, J. J. Green, and Pepin Tuma all read and improved the material. My expert consultants—Robert Phalen, Fred Lipfert, Roger McClellan, Neil Roth, Joseph Lyon, M.D., and Suresh Moolgavkar, M.D.—were helpful during the entire course of the project, from the day my research began to making improvements and finding errors in the near-finished work. That said, any remaining errors are my responsibility alone. Finally, I wish to thank the John M. Olin Foundation for its financial support.

1

Something in the Air

The EPA's new proposal, which became part of the National Ambient Air Quality Standards (NAAQS) of the Clean Air Act, would reduce the amounts of two types of pollution that regions would be allowed to tolerate.[1] This would mean increased restrictions on automobiles, factories, utilities, and many other things. The restrictions could range from installing new filters or other devices, to requiring the use of different types of fuels, to shutting businesses down entirely.

The first type of pollution comprises particles in the air, also called particulates. Particulate matter, often abbreviated as PM, can be solid or liquid. Some particulate matter can be seen in clouds, but individually most pieces can be viewed only with an electron microscope. PM can be emitted directly or can be formed in the atmosphere when gaseous pollutants (called precursors) react together. One common way of categorizing PM is by size. Thus PM10 more or less includes particles that are ten microns wide or smaller. (A micron is a millionth of a meter, 0.00004 inches, or one-hundredth the width of a human hair. Do not try to find this measurement on your ruler.)

Although the EPA currently regulates only PM10, the new standards also specifically regulate so-called fine particles, meaning those that are 2.5 microns wide or less, called PM2.5. Different areas of the country vary in their sources of pollution, but most PM2.5 comes from the precursors sulfur dioxide (mostly from power plants and manufacturing plants) and nitrogen oxide (mostly from power plants, manufacturing plants, and vehicles).[2] When they combine with oxygen they are transformed into fine particulates.

The second type of pollution the EPA is seeking to reduce further is ground-level ozone, better known as *photochemical smog*. Ozone is formed when two different precursor gases—first, volatile organic compounds (VOCs), mostly from vehicles, but also from such diverse sources as freshly applied paints and solvents, backyard barbecues, and numerous industrial processes, such as bread baking and dry cleaning clothes, and second, the aforementioned nitrogen oxide—mix together and are "cooked" by sunlight.

The EPA readily admits, and has published data that clearly show, that the levels of all these pollutants are already dropping in the United States (see table 1–1). So why the new action? It began when the American Lung Association (ALA) sued the EPA and won, on the grounds that the Clean Air Act requires the agency to review its air quality criteria every five years and the EPA had not done so for particulates. But all the ALA had earned was a review.[3] Nobody ordered the EPA to set a new standard, much less a tighter one. That was entirely the EPA's decision. It was also the agency's choice to propose a stricter standard for ozone, as well as for particulates. It claims to have done so because there are common factors between ozone and PM emissions, especially nitrogen dioxide. But the same could be said of many other types of emissions as well, as Jack Gibbons of the White House Office of Science and Technology Policy (OSTP) pointed out in urging that the ozone standard be delayed.[4] Critics say that the EPA's real motive is probably to confuse the issues, letting the strengths of some of its PM arguments mask the weaknesses of its ozone arguments and vice versa.[5]

TABLE 1–1
SUMMARY OF AIR QUALITY AND EMISSIONS TRENDS, 1986–1995
(percent)

Targets	Decrease in Concentrations	Decrease in Emissions
Carbon monoxide	37	16
Lead	78	32
Nitrogen dioxide	14	3
Ozone	6	a
PM10	22	17
Sulfur dioxide	37	18
VOCs	b	9

NOTE: PM10 measurements began in 1988.
a. Since ozone is not directly emitted, but rather forms from other emissions, it is not measured under this category.
b. Since VOCs are emitted as such but become other things in the atmosphere, they are not measured under this category.
SOURCE: EPA, *National Air Quality and Emissions Trends Report, 1995* (EPA, Office of Air Quality Planning and Standards: Research Triangle Park, N.C., October 1996), p.1.

Still, the Clean Air Act does require the EPA to determine that pollution threshold at which some health risk is posed, and then to set an air quality standard *below* it, allowing an "adequate margin of safety."[6]

If an area is not in compliance, then the agency can threaten to withhold highway or other funds, thereby forcing the local authorities to implement controls on local businesses and citizens. "The EPA is more like the Don, making local governments the hit men," says Bonner Cohen, the editor of the *EPA Watch* newsletter in Washington, D.C. The hit men, in turn, are supposed to make local businesses and vehicle owners an offer they can't refuse. In southern California, air pollution regulations have had a devastating effect on industry and have forced many companies to flee to other states or to Mexico.[7] Elsewhere, it is difficult to say how much of this has happened so far. But as the California example shows, businesses can be ordered to change practices or move, car owners can be made to add extra equipment, and new building and expansion permits can be denied.

Back in 1994, EPA regional director Peter Kostmayer threatened, in the words of the *Pittsburgh-Post Gazette,* to "immediately impose sanctions that could crush economic growth" in Pennsylvania if the state did not go along with the EPA scheme to build eighty-six centralized automobile emissions test centers, as required under the Clean Air Act. For example, he said, any firm that obtained permission to emit air pollution could do so only if the state could find another business willing to cut pollutants by twice as much. He also said that the state would lose about $1 billion a year in federal highway funds.[8] Later the EPA said, in essence, "Whoops, we guess you didn't need those test centers after all." Too late. They were already built. So under threat of a lawsuit, Pennsylvania had to pay the test center company, Envirotest, $145 million for a few score of empty buildings.[9]

The Particulate Hunters

The EPA originally claimed the new standard for fine particles would prevent 20,000 premature deaths annually,[10] though it later lowered that estimate to 15,000.[11] That's a lot of deaths, although less than seventh-tenths of a percent of the 2.2 million deaths of persons of all ages each year in this country. The EPA also claimed the new standard would prevent 250,000 occurrences of respiratory symptoms in children and 250,000 occurrences of aggravated asthma each year.[12] But the Natural Resources Defense Council (NRDC) claims, "Approximately 64,000 people may die prematurely from heart and lung disease each year due to particulate air pollution," and all but about

8,000 of these could be prevented by adopting even tighter standards than the EPA has proposed.[13] The NRDC, instigator of the Alar scare in 1989, is probably the nation's most effective environmental group.[14] The ALA has gone a step further, demanding PM2.5 standards that are even tougher than those proposed by the NRDC.[15] Unlike the NRDC, the ALA does not propose an alternative death rate, but it does claim that the EPA particulate standard would leave more than 28 million Americans "at risk" of lung and heart diseases.[16] The NRDC and ALA are also pushing the EPA to make the PM10 standards tougher.[17]

The NRDC, in a 1996 report titled *Breath-Taking*, tells us in large italicized words placed in the margin, "By the late 1970s it was widely recognized that smaller, inhalable particles were more significant to human health than relatively large particles."[18] But if this were so widely recognized, then surely the NRDC should be able to offer a better authority for the source than a book coauthored by Frederica Perera, a virulently anti-industry scientist who was working for the NRDC at the time.[19] In fact, there is precious little data on PM2.5 even today, and there was far less in the 1970s. Although the EPA first started regulating particulates in 1971, it was not until 1987 that the agency itself switched to regulating just those particles of PM10 or smaller.[20]

The current PM10 standard allows an annual average concentration of 50 micrograms per cubic meter of air. It also allows a maximum of 150 such micrograms in a twenty-four-hour period. Nevertheless, concern has grown that among the PM10 particles, the worst of the worst are those that are PM2.5 or smaller.[21] These are the most likely to slip past nose hairs, mucous in the throat, the pharynx, bronchial tubes, and other bodily defenses and to lodge deeply in the lungs—although eventually they do get expelled or are dissolved. Speaking at a June 26, 1997, press conference with Carol Browner standing at her side, Director of the Office of Environmental Quality Kathleen McGinty told reporters, "The president today is taking a major step forward with regard to a new pollutant that has never before been controlled. That pollutant is very small tiny [sic] particles that we know cause premature deaths."[22] This isn't true. By regulating PM10, the EPA to an extent already regulates PM2.5, because it is a subset—in the same sense that regulations on all cars are regulations on small cars. Indeed, very roughly speaking, if you collected a pound of PM10, about eight ounces of that would be PM2.5, although this varies tremendously from town to town and region to region. But specifically regulating PM2.5 would put more focus on both monitoring and eliminating those particular particles. The new standard would allow no more than a yearly average of 15 micrograms of PM2.5 per cubic meter of air, with a maximum twenty-four-hour average of 65 micrograms (see table 1–2).

By everybody's estimation, the requirements of the new EPA standard (much less the ALA's or the NRDC's) are going to be a lot harder to meet. One hundred twenty-six counties that comply with the agency's current particulates standard would fail to meet the new ones—"noncompliance" counties would grow fourfold, from 41 at present to 167.[23]

Most *primary* PM2.5 comes from dust, soil, and wood burning,[24] leading some opponents of the EPA's proposal to state erroneously that these—and not the usual suspects of cars, manufacturing plants, and utilities—are the greatest contributors to PM2.5 in the air.[25] But if the precursor pollutants are included, the major source of the offenders is the burning of fossil fuels, especially coal.[26] Petroleum products and, to a lesser extent, natural gas are also important contributors. So the EPA is, indeed, rounding up the usual suspects of cars, manufacturing plants, and utilities. To get fewer of these fine particles means burning less fuel or burning it more cleanly.

The EPA, the environmental groups, and some air pollution researchers would have us believe that, without a doubt, particulates are a proven and serious cause of ill-

TABLE 1–2
CURRENT, NEW, AND RECOMMENDED AMBIENT AIR QUALITY STANDARDS FOR PM10 AND PM2.5 MICROGRAMS PER CUBIC METER OF AIR

	Current EPA	New EPA	Recommended NRDC	Recommended ALA
For PM10				
24-hour average	150	150	33	50
Annual average	50	50	17	30
For PM2.5				
24-hour average	None	65	20	18
Annual average	None	15	10	10

NOTE: New standards will not take effect until after the year 2000.

SOURCE: EPA, *Review of the National Ambient Air Quality Standards for Particulate Matter,* pp. VII-46–47; ALA, "Comments of the American Lung Association," p. 2; NRDC, Deborah Sheiman Shprentz et al., *Breath-Taking: Premature Mortality Due to Particulate Air Pollution in 239 American Cities* (N.Y.: National Resources Defense Council, 1996), pp. 129–30.

ness in the United States at current levels in many of our cities. The smaller particles are the most egregiously harmful, they tell us, but particles in general are Very Bad Things. They speak loftily in terms of a "fairly consistent and robust relationship," as one national health official phrased it, between particles and health problems ranging from coughing to death.[27] Carol Browner is not only on the side of the angels, she says, but on the side of solid science. "Both for ozone and particulate pollution," she said at her November 27, 1996, press conference, "the scientific findings are clear. The question is not one of science."[28] In testimony before Congress in February, Browner said that the EPA's Clean Air Scientific Advisory Committee (CASAC) had reviewed "eighty-six studies . . . indicating that our current air standards are not adequately protecting the public's health."[29] On other occasions, she has said eighty-seven. [30] (Actually, it is eighty-six; but whatever.) She repeats it over and again, like a mantra. I listened to this repetition at one congressional hearing and was tempted to bump her, to get the needle to jump to the next groove on the record.

In fact, most of the major studies were conducted by a handful of researchers, only one of whom has a medical background and one of whom formerly worked for the EPA and has strong environmentalist ties. Moreover, a large number of highly respected scientists do question the whole PM2.5 and even PM10 paradigm. They are not kooks, industry dupes, or Elvis sighters. They include medical doctors and highly trained epidemiologists. They publish in major, peer-reviewed medical and scientific journals. And persistently, the EPA, the environmentalists, and the media simply ignore them. Why? Because they reach the "wrong" conclusions. Consider the shootout in Utah Valley.

2

How Grim Was My Valley?

In Utah Valley, in the north central part of the state, on a clear day you can see, well, pretty darned far. But on days when there is an air inversion caused by the trapping effects of the Wasatch Range to the west, a yellowish brown haze covers much of the sky. "It was a stunning effect," says one former resident; "almost as if somebody put a lid above the top of a trash can fire."[1] If air pollution at current levels were killing Americans, you would expect it to be happening in Utah Valley. Further, you would expect it to be relatively easy to detect because of another unusual characteristic of the area—the population is 90 percent Mormon, and Mormons are forbidden to smoke. Indeed, survey data indicate that only about 5 percent of the people in Utah County are smokers,[2] as compared with about a fourth of the U.S. population as a whole.[3] Sure enough, an enterprising researcher from Brigham Young University did find a correlation between air pollution and health in Utah Valley. And though his findings were later refuted, the myth lives on.

C. Arden Pope, an economist at Brigham Young University in Provo, Utah, made a big splash both in the scientific community and in the popular media with an article in the prestigious *American Journal of Public Health* in 1989. His study endeavored to detect health effects from a shut-down (caused by a strike) of the Geneva Steel Mill in Utah County, which is in Utah Valley. This single steel plant, according to Pope, contributed between 50 and 80 percent of all the particulate pollution in the ambient air. With such an overwhelmingly nonsmoking population and such a dramatic discrepancy between excellent and bad air quality, the mill shutdown provided a singular opportunity to see just what particulate pollution can do to people.

Pope's conclusion: "PM10 levels were strongly correlated with hospital admissions," especially of children. Indeed, reported Pope, "children's admissions were two to three times higher" when the mill was open than after it closed."[4]

Pope's Utah work is the bedrock of the EPA-environmentalist position. But it turns out that the explanation for Pope's original findings had nothing to do with particulates, but instead with a contagious disease. "Every other year the Utah Valley has an epidemic of viral bronchiolitis, an infection of the tiniest tubes in the lungs," explains Joseph Lyon, M.D., a professor of epidemiology at the University of Utah. "It raises hospitalization rates dramatically. The year when the steel mill was closed was a low year for this disease."[5] Data that he and others presented in the January 1996 *Journal of Pediatrics* show exactly that. During the epidemic years, children's hospitalization rates for respiratory problems increased by 250 percent above those during the nonepidemic years.[6] These figures correlate exactly with Pope's finding of children's admissions being "two to three times" higher when the mill was open. The plant just happened to have been closed during a nonepidemic year.

Three years later Pope, along with Michael Ransom, his Brigham Young colleague, and Joel Schwartz, a former EPA researcher who now works at Harvard, looked for a correlation between high PM10 levels in Utah Valley and higher death rates. They claimed to have found a smoking gun, with the smoke comprising particulates. Although in sheer numbers the increases were small, since on average only two or three people die in Utah County each day, the highest particulate jumps, they said, led to a 16 percent increase in deaths.[7]

Could other researchers find these same results? Joseph Lyon and his colleagues at the University of Utah Medical Center used Utah County data for six consecutive years and found that in four of them, there was indeed a correlation between high PM10 levels and more deaths and hospitalizations for respiratory diseases, just as Pope and Schwartz had found. But for two of those years, there was no correlation. Why would PM10 kill and injure during four years, but not during the other two? In another study, published in the journal *Inhalation Toxicology*, Lyon and colleagues found no statistically significant increase in respiratory disease–related deaths following increased levels of PM10 in the air for *any* of eight years studied. ("Statistically significant," under the usual standard, means there is a 95 percent chance that the result was not caused by sheer chance.) They did find such an increase for two of the eight years for cardiovascular disease, but not for the other six.[8]

Then Lyon and his colleagues looked at adjacent Salt Lake County, which has an inversion problem they call "virtually identical" to that of Utah County. Salt Lake County has a large copper smelting plant that emits even more particulates than does Geneva Steel. But once again they concluded, "We essentially found no association in Salt Lake County to PM10."[9]

Lyon says that "if the relationship is causal" between particulates and hospitalizations, then "you expect it to be pretty consistent. So just how causal can this thing [particulates and hospitalizations] be?"[10]

Outside research has confirmed much of Lyon and his colleagues' work. Among the evidence is that of Patricia Styer and others of the National Institute of Statistical Sciences (NISS) in Research Triangle Park, North Carolina, who looked at both Salt Lake County and Cook County, Illinois, which includes Chicago. "The reported effects of particulates on mortality are unconfirmed," they concluded in *Environmental Health Perspectives* in 1995.[11]

Nonetheless, to the EPA, environmental groups, and the media, Pope remains infallible. With Joel Schwartz and another Harvard researcher, Douglas Dockery, Pope has created the Particle Hunter Triumvirate. Indeed, a great problem with the "science" behind the EPA's newest particulate regulation is that a preponderance of the American studies that find particles causing health damage were written by these three men. Further, the most influential of them, Schwartz, not only worked for the EPA but has allied himself with the NRDC. None of the three has a medical degree; none was trained as an epidemiologist. Pope is an economist, Schwartz a physicist. The only one trained in medicine is Dockery, who got his Doctor in Science degree from Harvard.[12]

This hardly puts them in the same category as the butcher, the baker, and the candlestick maker, but many scientists who disagree with their views believe that their lack of formal training is evident in their work. Add a fourth researcher, Lucas Neas, who works with Schwartz and Dockery at Harvard, and you have just accounted for another considerable portion of the literature. They have a very exclusive club going. They get lots of grant money, publish lots of papers, and almost invariably find positive correlations between particulates and sickness. But time and again, when other researchers attempt painstakingly to verify their results, they find they cannot do so.

3

Little Things Mean a Lot

Consider the case of Birmingham, Alabama. Joel Schwartz looked at health and pollution data from this city and found that deaths went up when PM10 levels went up.[1] In a separate paper he linked increases in PM10 with hospital admissions among the elderly.[2] Jonathan Samet of Johns Hopkins University, who does have medical credentials, performed an evaluation of the studies for the Health Effects Institute (HEI) in Cambridge, Massachusetts—a research group that the EPA and industry jointly fund—and confirmed Schwartz's findings.[3] Thus we are presented with three different papers showing that PM10 levels in present-day Birmingham are high enough to cause harm, even death. Case closed, right? Not exactly.

It is widely acknowledged that swings in temperature, either up or down, can hospitalize and kill people.[4] For this reason, particulate studies should take temperature into account. And the aforementioned Birmingham studies did so. But they did not account for one of Birmingham's most distinctive features—humidity. During the summer months it wraps you in a hot, sticky blanket of moist air and practically squeezes your breath out.

Researchers at the National Institute of Statistical Science in North Carolina, an area well known for its own humidity, did account for this factor when they carried out a follow-up study of Birmingham. In an as-yet unpublished report funded by the EPA, when NISS factored in changes of humidity, it found that both for illness and for death "the PM10 effect is not statistically significant."[5]

The Relative Importance of Relative Risks

It may be startling to hear that one small factor such as humidity can invalidate the entire results of a study. But it is less startling when we recognize that consistently, these particulate studies are anything but "robust." When researchers do find an increase in deaths or illness, the increase is generally so slight that a single unaccounted-for factor, such as humidity, can throw the whole study off. "The relative increase in total mortality and morbidity [sickness] associated with a 50 percent increase in air particulates is not large," admitted Clear Air Scientific Advisory Committee member and University of North Carolina epidemiologist Carl Shy, in his congressional testimony backing the EPA's proposed standard. He testified that it is "on the order of 5 to 10 percent above [that] of days with the lowest concentrations." But, he added, because this increase was spread across such a large population, we could be talking about a lot of sick and dying people.[6]

True, but that misses the point. When you have a minuscule increase, any slight error in your methodology can completely invalidate your results. Consider a horse race in which the winning steed placed first by the mere length of his foaming nostrils. One tiny factor, such as a bit of a stumble by the number two horse, could have made all the difference. Had the horse won by three lengths, it is a fairly good bet that he was truly the faster animal. In the particulate studies, time and again the Particulate Hunters' horse wins by just that nostril, or at most a nose.

When it serves their purpose, the public health community eagerly points out that a 5 or 10 percent increase in risk, or even a larger one, means little or nothing. Thus after a major study suggested that women who had had induced abortions suffered a 50 percent increase in breast cancer incidence, an editorial in the prestigious *Journal of the National Cancer Institute* declared, "A typical difference in risk (50 percent) is small in epidemiological terms and severely challenges our ability to distinguish whether it reflects cause and effect or whether it simply reflects bias."[7] The American Cancer Society's Eugenia Calle added, "Epidemiological studies in general are not able, realistically, to identify with any confidence any relative risks lower than [a 30 percent increase.] In that context [a 50 percent increase] is a modest elevation compared to some other risk factors that we know cause disease."[8] Meanwhile, the National Cancer Institute declared that increased risks of less than 100 percent "are considered small and are usually difficult to interpret [and] may be due to chance, statistical bias, or effect of confounding factors that are sometimes not evident."[9]

So with abortion a 50 percent increased risk means nothing, we're told, while with particulates a 5–10 percent increase provides enough confidence to foist scores of billions of dollars worth of new regulations each year on the public. The only instance when a federal agency has suggested using anything near such a tiny apparent increase as a basis for regulations was—yes, the EPA. That was when the agency decided that a 17 percent increased risk provided solid enough evidence to institute sweeping regulations against environmental (passive) cigarette smoke.[10]

In fact, the particulate studies provide a wonderful example of *why* small increases in risk factors may mean nothing. It need not be a mistake or an oversight that converts the results of a study from positive to negative—the selection of a different period of time is sufficient. The Birmingham case is an example. Neil Roth of Roth Associates in Rockville, Maryland, looked for a pollution-health hazard connection in Birmingham using later data than Schwartz used. "Ours went from the late '80s into the early '90s, while his was from 1985 through 1988," says Roth. "We did thousands of different analyses on both the hospital admissions and the mortality data," he says, and "found the overwhelming majority of results were not statistically significant. Of those that were, half were positive but the other half were negative. So in short, we found no evidence of health hazards from particulates in Birmingham."[11]

This happens time and again. In one instance, Schwartz turned over to a group of statisticians at Stanford data that he had collected concerning his study finding a significant association between hospital admissions and particulate pollution in Detroit.[12] The statisticians found that when they analyzed them in the same way he did, they arrived at the same finding. But when they incorporated the potential influence of the day of the week into the model (that is, the fact that people tend to go to hospitals on some days, especially Mondays, rather than on others), particulate matter was no longer significant.[13] Similarly, when Suresh Moolgavkar, M.D., of the Fred Hutchinson Cancer Research Center in Seattle, Washington, and other researchers looked at Schwartz and Dockery's work correlating deaths with particulates in Steubenville, Ohio, one of America's most polluted cities,[14] they found no such connection.[15] When Schwartz looked at Minneapolis-St. Paul, Minnesota, he found that particulate increases caused more old people to go to hospitals.[16] When Moolgavkar and associates looked, they found the strongest association to be with ozone. Particulates did at first glance appear to be associated with admissions, but once gaseous pollutants were included in the mixture, the association with particles disappeared.[17]

The Philadelphia Story

The City of Brotherly Love has also proven a lovely city for persons studying the health effects of particulates. Again,

Joel Schwartz and Douglas Dockery found that as particulates went up, deaths went up.[18] Three years later, in 1995, Neil Roth and his colleague Yuanzhanh Li did a follow-up study and found no connection between particulate increases and deaths.[19] So did Moolgavkar and associates, that same year. Looking at a number of types of pollutants, they concluded, "No specific pollutant can be singled out as being responsible for the association between air pollution and mortality." Instead, they said, writing in the journal *Epidemiology*, "The particulate component of air pollution appears to have become the villain because it is a ubiquitous component of air pollution and thus serves as a proxy measure of pollution."[20] Jonathan Samet concurred with Moolgavkar's judgment in an editorial in the same issue.[21]

Despite all this, Carol Browner speaks of the "consistency and coherence" of the studies on particulates.[22] If there is any consistency at all, it appears to be that the Particle Hunter Triumvirate consistently discovers that particles are unhealthy, and other researchers are consistently unable to find these same effects when they look at the same cities.

Here enters the important concept of particulates as a proxy. Does this mean they have the power to vote for others at shareholder meetings? No, Moolgavkar is making a point that is expressed regularly by those who are skeptical about the health hazards of particulates: pollutants tend to accompany other types of pollutants. There are two reasons for this.

First, rarely does something that emits any pollution into the air emit only one type. Any combustion of fossil fuels will throw a whole host of things into the air, whether the combustion is of oil, coal, or natural gas, whether from a car, an electric utility, or a manufacturing plant. A gasoline-burning automobile engine, for example, emits fine particles but also nitrogen, carbon dioxide, water, carbon monoxide, several hundred types of volatile organic compounds, unburnt fuel, nitric oxide, nitrous oxide, and traces of other things.

Second, when particulate levels rise or fall, the levels of other types of pollution tend to rise and fall as well. Exhaust from utilities and other fuel-burning plants will be much the same for each Monday (barring holidays), each Tuesday, and so forth. What *will* change is what happens to those emissions once they are in the atmosphere. Will they dissipate and blow away, or will they just hang there and concentrate?

If a city suffers an inversion and stagnant air, the density of particles will go up, but so will lots of other things. Some will be man-made; others will be natural, like dust and pollen. "Everything goes up and goes down at the same time," says Fred Rueter, vice president of the Consad public policy research group in Pittsburgh and an adjunct professor at Carnegie Mellon University.[23] Some of those things can be and are independently measured. Others are not looked for or cannot be looked for because of a lack of monitors. The result is that particles may be just a proxy, or a marker or a surrogate if you prefer, for something else in the air. "If one person who dropped dead had consumed sugar laced with strychnine," says Moolgavkar, "and another who also dropped dead had consumed sugar laced with potassium cyanide, would we blame the sugar?"[24]

Moolgavkar and his associates note that some have "argued that the association of particulates with mortality is remarkably consistent from city to city, in the presence or absence of other pollutants, and under varying conditions of weather." But, they say, they have "been unable to identify a single study in which other pollution variables have been adequately controlled."[25] That was in 1995. Does Moolgavkar still feel the same way in 1997? Yes, he told me. "When you look at the studies, you generally see that only one pollutant is observed at a time. They [the particle pursuers] don't look at the complex mixture; they just focus on particulates. Often times they have the data available, but they just don't use it."[26]

4

A Tale of Six Cities

The aforementioned Steubenville, Ohio, also played a major role in an ambitious study by Douglas Dockery, Arden Pope, and others comparing air pollution and premature deaths in six American cities, the so-called Harvard Six Cities study, published in the December 9, 1993, *New England Journal of Medicine.* Along with one other study that will be discussed in the following chapter, based on American Cancer Society health data and called ACS II, the Six Cities study is by far the most important weapon in the EPA's PM2.5 advocacy arsenal. In it, the researchers found that Steubenville, with the most air pollution and the most particulate pollution of the six cities, had a 26 percent higher mortality rate than had Portage, Wisconsin, the cleanest city.[1]

But what they also found, and what none of the particulate pursuers ever talk about, was that among nonsmokers there was no statistically significant increase in deaths between Steubenville and Portage. Moreover, there was none if you excluded persons with occupational exposures to "gases, fumes, or dust." It was only by including the smokers, the former smokers, and the persons with occupational exposures that they were able to get significant findings. It's right there in table 3 of the study.[2] But in the text,

Dockery and his associates state that "although cigarette smoking and other risk factors were associated with mortality, our estimates of pollution-related mortality were not significantly affected by the inclusion or exclusion of these variables in the models." How do they figure that? Simple. In table 4, they abandon the "nonsmokers" category, replacing it with a new one that blurs the distinction between the "nonsmoker," "current smoker," and "former smoker" categories.[3] Voila! They have now converted the entire Six Cities study from negative to positive. Do that in your doctoral dissertation and it can get you thrown out of school. Do it in a medical journal and you become the hero of the EPA and the environmental community.

Even if they had found significance as they claimed, it would be a very weak study. While comparing one city with another may seem a nifty idea, it can cause even more headaches in terms of getting comparisons just right. Now you are not only adjusting for temperature, humidity, and a few other variables, you are also trying to adjust for every health difference between each city. You do not want to make the mistake of concluding, say, that the reason Denver has a lower heart disease rate than Chicago is that Denver has thinner air, more sunshine, prettier views, or a worse professional basketball team.

Critics insist that Dockery and Pope did not make those proper adjustments. For example, not only did they not control for humidity, they did not even control for temperature. Yet temperature extremes have been associated with a 30 percent increased chance of death.[4] Nor did they consider income differences between cities. Together, these two could easily throw off the whole study. That is because Steubenville is considerably poorer than Portage.[5]

"Poor persons tend to die more quickly during extreme weather conditions than wealthier ones," says Roger McClellan, a former Clean Air Scientific Advisory Committee chairman and president of the Chemical Industry Institute of Toxicology in Research Triangle Park, North

Carolina.[6] As part of the National Health Survey, a 1996 study found that persons living below the poverty line, for a variety of reasons that include unhealthier lifestyles and habits and less medical screening, are far more likely to become sick than are wealthier persons. Persons covered by the federal insurance program for the poor, Medicaid, were four to five times more likely to have emphysema, chronic bronchitis, and congestive heart failure than were those with incomes above the Medicaid qualification level.[7]

Meanwhile, Fred Lipfert, a scientist at the Brookhaven National Laboratory in Upton, New York, who has been studying air pollution since the 1970s, has done an analysis in which he has found a correlation between sedentary lifestyle and premature death in five of the six cities.[8] "This can explain the variation that Joel Schwartz and his fellow researchers had attributed to air pollution," says Lipfert.[9]

Looking for Pollutants in All the Wrong Places?

Many skeptics believe that the research revolving around the Schwartz-Dockery-Pope axis is self-fulfilling. In a paper in the *Journal of the Air and Waste Management Association*, Lipfert and Ronald Wyzga, senior manager of Air Quality Risk and Health Assessment at the industry-funded Electric Power Research Institute (EPRI) in Palo Alto, California, found that if a research team had chosen to focus on sulfur dioxide or nitrogen dioxide instead of on particulates, their work would have found the same effects on daily deaths. A focus on carbon monoxide showed somewhat larger effects than were shown by particulate matter, even though practically nobody argues that carbon monoxide— except at the massive doses you get from a leaky heater or auto exhaust in an enclosed garage—is fatal.[10] "An obvious conclusion," says Wyzga, "is that if an investigator had elected to study another pollutant instead of particulate pollution, he or she might well have concluded that the other pollutant was the pollutant of concern."[11]

But how can this be? Is it that all these pollutants are causing health problems? Maybe, but probably not. A more probable answer is that *none* of them is; that when weather conditions increase the volume of pollutants, more people die because of the weather fluctuation itself, not because there is more of this stuff in the air.

It is fairly common for air pollution epidemiological studies to control for temperature fluctuations. Considered far less often is air movement. Yet "these are the main cause in day-to-day changes in ambient concentrations of pollution," notes Fred Rueter.[12]

Thus particulate levels—or those of carbon monoxide or sulfur dioxide or nitrogen dioxide—are all markers for stagnant air. So what's the point here, that stagnant air is the real killer? No, not directly. But it has long been accepted that pollution is normally much worse inside most homes than directly outside. Indeed, the average American *adult* spends 93 percent of his time indoors.[13] This makes indoor air far more important than outdoor air. And, notes Rueter, "when particulates rise, so does almost everything else, and that includes indoors as well as outdoors."[14]

That is because stagnant air causes the exchange rate between outdoor and indoor air to go down. On those occasions when air becomes stagnant, pollution concentrations in indoor air increase as ventilation decreases.

These problems were exacerbated after the energy price shocks of the 1970s, when old homes and buildings were given more insulation and new ones were built with smaller windows, windows that didn't open, tighter seals, and more insulation. With less fresh air circulating into the house, indoor pollutants such as cooking byproducts, cleaning agents, mold spores, pet dander, tobacco smoke, hair spray, and cockroach and dust mite droppings would increase. "The notion that you're safer indoors than outdoors I think is problematic," says Lipfert.[15]

Regardless of how well or poorly a house or building is "buttoned up," the most important factor determining how much fresh air circulates into or out of it is how well the air is circulating outside. A windy day brings more fresh air in; a day with little wind brings less. Thus, the same conditions that allow pollution to build up outside allow it to build up inside.

"What I think is the most coherent explanation for all the epidemiological evidence associating outdoor particulate levels to illness," says Fred Rueter of Consad, "is that on the days when particulates are high, people indoors are being exposed to allergens producing effects ranging from mortality from asthma and other things to simple reductions in lung function, leading to increased complaints and hospitalizations."[16]

Another alternative that needs examining, according to Robert Phalen, a biomedical scientist who for twenty-two years has directed the Air Pollution Health Effects Laboratory at the University of California in Irvine, is that on days when the weather or some other factor causes people to spend more time indoors, by virtue of being inside they make certain outside pollutants increase. When they are inside, he says, they "use more heat, more air conditioning, and so forth, and this could cause several types of particulates outdoors to go up."[17] Thus the increase in particulates outside would simply be a marker for the hazard of breathing indoor air.

Harvard's Hide-and-Seek

None of this argument is presented in order to allege that the particulate hunters are dishonest. But under the old science, a researcher established a hypothesis and then tried to disprove it. Increasingly today, this method is being abandoned in favor of trying to prove a hypothesis because it either fits the researcher's political mindset, matches the desires of his funder, or supports positions on which he has built a reputation. Consider the police officer who is convinced that men with tattoos are more likely to commit crimes. By always keeping on the lookout for them and watching their actions, while ignoring people without tattoos, he is quite likely to verify his prejudice.

The material that Joel Schwartz and Douglas Dockery have in their database should be made public, rather than simply disseminated to persons they select. Yet for years they have refused to make it public—or even to give it to the EPA, although the federal government, including the EPA, paid for it. It isn't as if nobody has tried to get it. In early 1994, the Clean Air Scientific Advisory Committee wrote to Carol Browner asking that she get "crucial data sets linking exposure to particulate matter and health responses."[18] Several groups filed Freedom of Information Act requests with the EPA to get the Six Cities study data. The EPA had to respond that it did not have them.[19]

Finally, the Harvard researchers first announced they were going to give the pollution data to the Health Effects Institute (HEI), while withholding the health data—the information on hospitalizations and deaths. Later, under continuing pressure, Schwartz and Dockery said they would turn over the health data as well, albeit under certain stringent conditions.[20] The major problem now is that according to HEI President Dan Greenbaum, it will take two to three years to complete its evaluation.[21]

But the EPA has hardly been vigilant about trying to get the data. Mary Nichols, head of the EPA Office of Air and Radiation and a former director of the National Resources Defense Council's Los Angeles office, penned a letter to Commerce Committee Chairman Thomas Bliley (R-Va.) saying there was no need "for EPA to obtain the underlying data," since the studies were published in peer reviewed journals.[22] Browner later told Congress the same thing.[23] This misses the point; a reviewer cannot review what he cannot see. All he can do is to ensure that what is before him is correct. If the author is selectively presenting data that fits his hypothesis, there is precious little a medical journal or peer reviewer can do.

24

Greenbaum says he sympathizes to some extent with Schwartz and Dockery's recalcitrance. "It's a complex database," he says, "and you need procedures to make sure only qualified investigators can look at this stuff. Some people tend to be analysts for hire. You need to have a thoughtful analysis of it." Certainly you do not just want to post it on the Internet, where somebody like Pierre Salinger can use it to claim that fine particles were responsible for the downing of TWA Flight 800.

But Schwartz's motives seem a bit darker when he says that he does not want to provide the information to "industry thugs" (as he put it to the *Wall Street Journal*), by which he means scientists who receive considerable industry money.[24] But by allowing only selective access to the data, he may be trying to ensure that nobody can have access to the data who may reach the "wrong" conclusions, with wrong being defined by him. "Any critical scientific inquiry is potentially subversive or blasphemous," observes University of Tübingen clinical psychologist Stuart Brody in a forthcoming book, "and the uncertainty required by such inquiry is itself unbearable for dogmatic persons, regardless of their political leanings."[25]

Indeed, some say Schwartz's prime motive may well be sweeping away his tracks. "He appears to pick out positive significant results that link air pollution to health problems," says Neil Roth. "Going back through Schwartz's results, we found he didn't report everything, only positive results"; that is, those linking pollution to health problems. Further, says Roth, "He always uses a different analysis in different cities."

For example? "Well, one factor is how many days of weather you're going to count before the day of death or hospital admission—the previous five days, the previous six days, or whatever. When he goes to different cities, he uses a different lag time," Roth says. "Had he been consistent he would have found insignificant results in some of his cities."[26]

During February congressional testimony, Sen. Jeff Sessions (R-Ala.) asked Schwartz if he had calculated for humidity in Birmingham. Schwartz admitted, "In that study, no. Frankly, I haven't seen humidity being put in lots of other studies."[27] Yet the EPA staff paper on particulates asserts, "Most [short-term PM studies] include temperature and dewpoint as covariates in their studies."[28] Dewpoint is a measure of humidity.

Suresh Moolgavkar at the Fred Hutchinson Cancer Research Center says he also sees such problems in Schwartz's work. But it is not just Schwartz who does this, he says. "If you look at various PM studies, some will use a lag of two days, some one day, some even three or four days. To me that's not consistency. To me, it means you have to go through contortions to get a specific result."[29]

Protective Polish Particles?

Although they are behind the times by American standards, the Europeans are now studying the particulate issue. In 1996, the *Journal of Epidemiology and Community Health* published data from fifteen cities in ten different European countries to try to detect associations between various air pollutants and increases either in deaths or in hospital admissions. None of the studies looked at fine particles directly. They studied all particles, or in some cases they simply omitted from consideration the largest ones. The only consistency among the study findings is that they were completely inconsistent.

A statistically significant increase in death or illness was found in London, England;[30] Paris, France;[31] Barcelona, Spain;[32] Athens, Greece;[33] and two of four Polish cities studied.[34] No statistically significant increase was found in Helsinki, Finland;[35] Lyon, France; Cologne, Germany;[36] Amsterdam and Rotterdam, the Netherlands;[37] Bratislava, the Slovak Republic;[38] and the other two of the four Polish cities studied. Indeed, in the city of Wroclaw, respiratory deaths went way down when particle levels went way up, and yes, the association was statistically significant.[39] In Milan, Italy, for persons under age sixty-four there was no statistically significant association for particles and hospital admissions, while for those over sixty-four, under some circumstances there was and under others there was not.[40] For arithmetically challenged readers, that makes six cities with positive associations, eight without, and one mixed.

Other work in Europe also shows mixed results. Studies in which Joel Schwartz has a hand seem to find that particulates are hazardous, while many of the rest do not. At an international conference in April in Prague in the Czech Republic, Neil Roth presented the results of his study about Prague itself. Again, there was no association between particulates and either hospital admissions or mortality. "This is especially important because pollution levels in Prague are so high compared with most American cities," says Roth. "It was a place where you might expect to find effects." Further, the Prague data, since they come from a country with socialized medicine, give more thorough hospital admissions information than is available in the United States. "They capture everything," he says, meaning they have records on everyone.[41]

All of which hardly seems to provide a case for the "remarkable consistency" that the particulate hunters like to talk about.

5

The Fine Particle Follies

Recall that originally the EPA regulation covered all particles, then it switched to PM10, and now it wants to emphasize PM2.5. Rueter says the best way of characterizing this is to say that "They're moving from PM10 because they can't demonstrate that what they've controlled to date helps, so it must be something else."[1] Undaunted, Carol Browner speaks with the greatest confidence that the third time will be a charm. "The scientific evidence indicates that very small particles pose the greatest risk to human health and are most likely to lead to respiratory complications, including death," she said in announcing the standards in November.[2] On an earlier occasion, the EPA administrator had said, "The smaller particles that penetrate farther into the human lung present the greatest risk." But that was not Browner; rather, it was William Ruckelshaus proposing the *PM10* standard in 1984.[3] So, here we go again.

Still, backing up Browner are the American Lung Association, the Natural Resources Defense Council, and researchers like Joel Schwartz. The ALA says that "there is a definite and undisputed correlation between any measure of PM2.5, whether direct or indirect, and mortality and morbidity effects," and that "the body of evidence supporting a focus on PM2.5 is compelling."[4] In a statement released with the NRDC report, Schwartz states that it "has become clear" that "any concentration of fine particles in the air is associated with increased risk of death, of hospital admissions for lung and heart disease, and of increased respiratory symptoms," and that "it is the fine particles, that is PM2.5 . . . that are the problem."[5]

Death and Witchcraft

The EPA has ostensibly compiled all the relevant data on PM into what is called the Criteria Document.[6] If you want to get a copy, bring a forklift—it runs to some 2,400 pages. But this poundage masks a vital fact; only a handful of the eighty-six studies contain direct measurements of fine particles. The rest either estimated fine particles as a percentage of the total or did not discuss fine particles at all. I contacted Harvard's Douglas Dockery for a list of all published studies that had direct PM2.5 measurements. He kindly sent one. As it happens, one of these did not look at particles,[7] and another was simply a measurement of particles in the atmosphere and not a medical study.[8] For some reason he left out one study of which he in fact had been a coauthor. The study found a positive association. Add to this a just-published study, and the number stands at a baker's dozen.

First, let's consider the studies looking for premature deaths. (See appendix A.) There were four of these, of which one used the database from one of the others. Thus there were only three datasets. Each of the four papers had as its lead author one of the members of the Particle Hunter Triumvirate—Joel Schwartz, Douglas Dockery, or Arden Pope. Nonetheless, only one unequivocally found a significant association between fine particles and premature death among nonsmokers and persons without occupational exposure. This was a 1995 study of which Pope was the lead author. It relied on American Cancer Society health data, and it is the aforementioned ACS II. It appeared in the *American Journal of Respiratory and Critical Care Medicine,* published by the American Thoracic Society, a component of the American Lung Association.[9] To some, this connection automatically makes the study suspect,[10] but we need not resort to such simple tarring to find real problems with it.

One such problem, says Suresh Moolgavkar, is that "they didn't adjust for such obvious things as weather changes." No, even temperature was not considered in the ACS II study. He adds, "There was a correlation between pollution and mortality, but they did not look at any pollutants other than sulfates and particles."[11] Again, Moolgavkar's case is supported: if fine particles are simply a marker for something else, that "something else" cannot possibly be detected if the analyst is not looking for it.

But probably the biggest problem with ACS II is that although the PM2.5 data were collected from 1979 to 1981, the death data covered the years 1983 through 1989.[12] "Did the PM2.5 level change after this?" asks Joseph Lyon at the University of Utah Medical Center. "Did they clean up the air? Arden [Pope] doesn't know and nobody else knows."[13] Pope and the other authors seem to be making the assumption that either the PM2.5 level remained the same in all the compared cities or that if the level changed, it changed equally in all the cities. Neither assumption could possibly be true; indeed, while Pope used Fred Lipfert's collection data, Lipfert himself in a 1995 paper noted that during this time period the dirtiest cities could have been expected to be cleaning up their air at a much quicker pace than were the cleaner cities, which were already in compliance with federal law.[14] This could make the particle data absolutely useless.

A third assumption would be that the deposition of fine particles in the lungs from 1979 to 1981 was causing deaths several years later. Certainly, when it comes to smoking and lung cancer, we know that there are extremely long lag times between exposure and diagnosis—often more than twenty years. But ACS II found no associations whatsoever between lung cancer and PM2.5 exposure. All the increases in death came from cardiopulmonary disease. [15] Furthermore, the argument that pollution levels were influencing deaths years later would provide no support for the PM10 studies—including several by Pope himself—that the EPA cites showing sudden spikes in particulates corresponding to increases in deaths a few days later.

The problem of linkage between pollution data collected during one period and health measurements made several years later is not insurmountable, and thus the criticism of it is hardly unfair. The health data are still being collected. Researchers must obtain the current PM2.5 statistics, rather than rely on those more than fourteen years old. Doing so, of course, takes money and time.

As weak as the ACS II study is, so little research is available that somehow it has become something of a "gold standard"—with consequences that were to embarrass the EPA in April 1997. By extrapolating from one table in this one study, the EPA concluded that fine particles were killing 20,000 Americans prematurely each year. But a critic of the proposed standard, Kay Jones of Zephyr Consulting in Seattle, Washington, found that Pope and his fellow researchers had made a simple miscalculation in that table. This forced the EPA to revise the 20,000 down to 15,000. [16] If only it were always so easy to save 5,000 lives!

The point is not to bust the EPA's or Pope's chops for making a mistake, or, as a Clinton aide said in a different context, that "mistakes were made." Rather it is to show what can happen when you put all your particles in one basket because that is the only basket you have.

Of the four premature-death PM2.5 studies, only one besides ACS II showed increased deaths among nonsmokers. It found statistically significant associations between particle increases and deaths in three cities, and no associations in the other three it looked at. Importantly, one of the negative cities, Steubenville, also had the largest increase in fine particles. This is just the opposite of what we would

31

expect to find if particulate increases were causing deaths. But in any case, because there was so much information from one of the significant cities (Boston), the association in all the cities combined became statistically significant.[17]

How about symptom, illness, and hospital admission studies? (See appendix B.) The American Lung Association published a report in 1996 called *Gambling with Public Health II* that complained that the EPA was being too soft on particulates even with its proposed standards. According to the report, "A number of studies have linked fine particle pollution (PM2.5) with increased prevalence of coughs and bronchitis, with especially severe effects on children with asthma."[18] Really? Of the nine PM2.5 studies looking at illness but not death, four showed *no* association between fine particles and hospitalizations or *any* of the symptoms measured. Five showed an association between fine particles and *some* of the symptoms measured.

"We've just arbitrarily decided PM2.5 is the villain," says Joseph Lyon. "This comes as close to witchcraft as anything I've seen. This isn't science. This is a firmly held warm and fuzzy belief: 'I know these things because I feel them.'"[19]

The problem is not simply that there are not many PM2.5 studies; there are not many PM2.5 *measurements* from which to build studies, given that only about fifty monitors capable of measuring such particles are operational at any time.[20] Says Lyon, "I would like to see some good studies done on PM2.5, but the data's just not there."[21]

Brookhaven National Laboratory's Fred Lipfert and the Electric Power Research Institute's Ronald Wyzga reviewed thirty published studies in the December 1995 *Journal of the Air and Waste Management Association*, which were all they could find, examining associations between deaths and particle pollution. Some of the studies used PM10 as the measure, others PM2.5. Lipfert and Wyzga then compared the two sets of studies. They concluded that PM10 was actually associated with a slightly *higher* risk, though there could not be a determination of statistical significance, since not enough studies were involved. Indeed, the EPA's own criteria document summarizes the results of the Harvard Six Cities study, including a table comparing the relative risk for increased death in adults. For PM10 or PM15, it is a mean increase of 42 percent. For PM2.5, it is 31 percent. [22] A mean increase is a comparison between zero pollution and the actual amount, as opposed to, say, comparing one polluted city with another. In other words, if you got rid of all the pollution, the health would change by that much. "In a subsequent analysis" presented at the April 1997 Prague conference, says Lipfert, "we have found that PM10 exhibits a higher mean effect [than PM2.5] for hospitalization and lung function as well."[23] Similarly, a study of Phoenix that was presented at a U.S. conference in April 1997 found a statistical association with deaths and PM10, but not with PM2.5. The chief author is a scientist with the EPA. [24] Although it has been widely circulated through *samizdat*, the EPA as of this writing has refused to release it.

Poster Prevarication

So how does the EPA deal with this damning evidence? Putting it pointedly, it patently prevaricates. Consider a poster that Carol Browner used in her February 12, 1997, testimony before the Senate Environment and Public Works Committee.[25] Entitled "Soot/Particulate Matter: The Science Calls for Action," it listed brief details of five studies involving PM2.5, along with the approximate number of persons involved, the alleged adverse health effects, and the average annual particulate matter concentration. (See appendix C.) Of these studies, only one clearly supports the "adverse health effect" the EPA claims—ACS II, with its myriad problems. And yet another problem has emerged with ACS II in this context, which is that it looked only at fine particles. So it was incapable of indicating whether other pollutants or larger particles might have been responsible. Another of the five listed studies is the one in which three of the cities showed associations and three did not. Browner's chart mentioned the three with positive associations. And the other cities? They were left off the chart.

Thus, the chart Browner displayed not only failed to show the range of the PM2.5 literature, making use of only five of thirteen studies, but of those studies it did mention it misrepresented the conclusions four out of five times. If there is a call to action here, it would seem to be disciplinary action against whatever EPA employee or employees prepared that chart for Browner. She herself would be guilty of perjury if she really knew whereof she spoke. But she has no fear of joining the infamous ranks of Alger Hiss and Aaron Burr. When questioned at a later hearing, she was forced to admit that she had read none of these studies, though she quickly added that her right-hand woman, Mary Nichols, had.[26]

Speculation, Hypotheses, and Just Plain Guesses

With such mixed results in the epidemiological data for both particles in general and PM2.5 in particular, with the positive results regularly failing verification, and with the positive results showing such slight increases, epidemiology will never give us the hard and fast answers we seek. "The epidemiological studies don't prove anything," says Daniel Menzel, a Clean Air Scientific Advisory Committee consultant and director of the department of community and environmental medicine at the University of California at Irvine (UCI). "They just show you there might be a relationship there."[27]

But that is no excuse for throwing our hands up in the air and declaring permanent ignorance. There are both clinical studies, that is, trials on human beings, and animal studies that can be done. Further, using animals with experiments involving lungs is not a practice fraught with the problems of animal carcinogen testing that I and others throughout the years have strongly criticized.[28] Rodent lungs are fundamentally like human lungs, and it is possible to test rodents with air pollutants at levels similar to what humans are exposed to.

The EPA Criteria Document does list some animal studies showing harm from fine particles,[29] but these have been essentially at concentrations far higher than the ambient concentrations correlated with human deaths in the epidemiological studies. Nobody denies that if you stick a rat's mouth where a vacuum-cleaner bag would normally go and then vacuum a dusty floor, you're asking for trouble.

But what about levels to which humans would be exposed on an exceptionally bad day in an exceptionally polluted city? To hear Joel Schwartz tell it, the animal studies show a clear and present danger. After stating that the epidemiological evidence against such particles was unequivocal, he told a congressional panel, "Recent animal studies have corroborated these findings, showing toxic effects of fine particles, especially in sick animals." The reason this is not widely acknowledged, he said, is that "industry has launched a lobbying offensive to convince political leaders and the general public with the same arguments that failed to sway a more technically sophisticated audience."[30]

Schwartz specifically cited the work of his fellow Harvard researcher John Godleski, saying it found that "particle concentrations that were not extraordinary, but comparable with concentrations seen in U.S. cities" had "killed rats with chronic bronchitis."[31] Actually, Godleski's study used a special device to concentrate fine particles into air so that the average fine particle level in the rats' chamber was 272 micrograms per cubic meter for six hours a day, three days in a row.[32] By comparison, the air in Steubenville, the dirtiest of the Harvard Six Cities, had an average of fewer than 30 micrograms of fine particles per cubic meter.[33] Indeed, Godleski and his fellow researchers wrote explicitly that the particle levels in the air given to the rats "was approximately thirty-fold" higher than it was outside their Cambridge, Massachusetts, window.[34]

That's one heck of a mistake for Schwartz to make. Meanwhile, UC Irvine's Menzel says, "To my knowledge there are no experimental animal data or controlled human studies which could explain epidemiological findings correlating particulate exposure to subsequent harm."[35] He says, "In my laboratory and that of my colleagues at UC Irvine we have found that experimental animals such as the rat are very insensitive to particulate matter exposures. We have never observed potencies equivalent to that proposed for humans based on the epidemiological data. This again raises the question of a plausible biological mechanism for action."[36]

As far as studies on humans are concerned, for ethical reasons they are limited in scope. But some have been done. As Fred Rueter characterizes them, "The studies on components of particulate matter considered the most likely culprits have found that people don't have measurable response to exposure anywhere near current concentrations. You need two to ten times the concentrations, and those tend to cause [just] minor respiratory problems." He says that "while studies do show that fine particles penetrate deeper into the lungs, what they don't show is that harm is done while this is happening."[37]

Another CASAC member, Mark Utell, M.D., along with Mark Frampton, M.D., both of the University of Rochester Medical Center, have concluded: "Available toxicological studies provide few clues in explaining acute mortality at low particle concentrations." Wrote the researchers, "controlled clinical studies with acidic particles at concentrations greater than twenty times ambient [levels] fail to produce [lung inflammation] in healthy individuals," and "subjects with [chronic obstructive pulmonary disease], the group at presumably highest risk [to judge from] the epidemiological data, show no reduction of lung function with similar acute exposures."[38]

The NRDC report dances gingerly around this problem of a lack of toxicological data. "At present, the toxicological mechanism or mechanisms associated with the observed increases in [death and illness] are not well understood," it states. It then sums up what is known, going researcher by researcher. From the very wording, it's clear that there's no *there* there, or at least not yet. Thus, says the NRDC, "Schwartz has suggested," "Ostro has suggested," and Schlesinger reviewed toxicological evidence that "might provide an explanation." Meanwhile, Oberdörster "postulates," while Seaton and colleagues have "proposed a hypothesis." The NRDC says Bates "hypothesizes," and Godleski has also "hypothesized." And that's it. The list comprises every researcher that the NRDC mentions in this section of the report.[39] Everybody has a hypothesis, a theory, a suggestion, a possible explanation. All of which is to say, none has yet found a shred of evidence.

Neither has the EPA. "The relevant toxicological and controlled human studies published to date have not identified an accepted mechanism that would explain how such relatively low concentrations of ambient PM might cause the health effects reported in the epidemiological literature," says the EPA staff paper. "An understanding of the biological mechanisms that could explain the reported associations has not yet emerged."[40]

How about the ALA? "The mechanism by which particulate matter affects mortality and morbidity is not currently well understood," states the *Gambling with Public Health II* report. "It is not known whether the size and concentration of the inhaled particles are wholly responsible for the health effects, or if their chemical composition also plays a role."[41]

University of British Columbia professor of medicine Sverre Vedal, M.D., presents the case much more strongly. "There is no known mechanism whereby exposure to very low concentrations of inhaled particles would produce such severe outcomes as death, even from respiratory disease, and certainly not from cardiovascular disease," he says.[42]

Ah, "causation, shmauzation," say the PM pursuers.

"It's like the debate over cigarette smoking years ago," says Morton Lippmann, professor of environmental medicine at the New York University Medical Center. "It is real, we just don't know why it is happening."[43] Joel Schwartz[44] and the American Public Health Association's Barry Levy[45] use the same comparison.

In fact, the cigarette example undermines their point. Ever since Europeans and North American colonialists began smoking tobacco, it has been regarded as harmful. The term "coffin nail" for a cigarette was coined a century ago.[46] That's because it was so terribly obvious that smoking killed, even if it was not exactly clear why it was so. Sure enough, epidemiological studies of smokers show they have a 1,500–2,000 percent increased risk of lung cancer. You can also use the "look-around-you" method to see that women over forty are more likely to get breast cancer than younger ones, or for that matter that blacks constitute a disproportionate number of professional basketball players and whites a disproportionate number of professional hockey players. All these observations can readily be made without precise statistics, without having to offer a good reason of why they are so. But there is no look-around-you factor with particulates. Nobody can actually name anyone who died of particulate inhalation; it's just a bunch of numbers built on speculation.

Unlike the activists and regulators, many air pollution specialists are not so willing to proceed without adequate information. "I think at this time, I would characterize the state of toxicological knowledge as interesting speculation," says Chemical Industry Institute of Toxicology (CIIT) President Roger McClellan. "But what we have to do in science is move from speculation and hypothetical statements to facts. I think we're just starting to get the data."[47] Robert Phalen represents many researchers when he says, "It is likely that when the research on particulate air pollution is more complete, the [PM2.5] standard will be demonstrated to be naive and unwarranted."[48]

6

The Asthma Anomaly—
Corporations or Cockroaches?

A panicked father rushes into the hospital with his gasping, asthmatic child in his arms. "Help!" cries the boy's mother. "He can't breathe!" The ad—sponsored by the Clean Air Trust in affiliation with the American Lung Association, Public Citizen, Defenders of Wildlife, and the Sierra Club—was part of a lobbying effort to support the EPA regulations.[1] So it is not hard to guess what the culprit is.

Taking their cue from President Clinton and his wife, who couch practically all their initiatives in terms of saving endangered children, proponents of the proposed EPA standards have done likewise. The "child card" is repeatedly played:

- "When it comes to protecting our kids, I will not be swayed," EPA's Carol Browner dramatically intoned at a recent conference on children's health in a push for the new pollution standards.[2]
- "Hundreds of scientific studies have shown that today's air pollution levels are shortening lives and harming children," claimed the National Resources Defense Council in a newspaper commentary.[3]
- The American Lung Association had young people with asthma testify at press conferences in support of the EPA standards.[4]

- The Sierra Club is running radio ads that use little children's voices to push the EPA regulations, saying how new laws will keep them from becoming sick.

Asthma is predominantly a childhood disease. Rates are indeed rising sharply among children—enough to merit the cover of the May 26, 1997, *Newsweek*, "The Scary Spread of Asthma."[5] At the June 26, 1997, press conference alongside Carol Browner, Kathleen McGinty told reporters that smog "seriously exacerbate[s] asthma conditions and other lung ailments and that particularly affect[s] children. Asthma is on the rise in the United States."[6]

Naturally, the environmentalists say, this rise is from air pollution, and only the white hats at the EPA can stop it. And who are the black hats? Syndicated *New York Times* columnist Bob Herbert asked readers to choose between "the kids with asthma who have a tough time breathing whenever there is a bad air day or the powerful representatives of the oil industry, the Association of International Automobile Manufacturers, the American Bus Association, the Chemical Manufacturers Association, etc."[7] We can all just picture some fat, cigar-chomping businessman sitting on the chest of a poor little child gasping for air. Asked specifically at the June 26 press conference, "Do you have estimates for how many counties will fall out of compliance?" Browner pleaded ignorance on this very sensitive issue. "We don't at this time. We are still looking at that," she said.[8]

The problem with that picture, though, is that as asthma incidence and deaths have been sharply rising, all the measured types of pollution—including particles and ozone—have been dropping. Furthermore, studies have failed to show a relationship between even high air pollution levels and asthma, as even *Newsweek*'s environmentalist writer Sharon Begley pointed out.[9] A recent comparison between asthma rates in highly polluted Leipzig in the former East Germany and the far cleaner Munich in West Germany found asthma rates lower in the east. Noting this and similar findings between squeaky-clean Sweden and polluted Poland, two researchers wrote in a January issue of *Science* that these findings "suggest that asthma prevalence has increased because of something lacking in the urban environment, rather than through the positive actions of some toxic factor."[10] Shortly before that, the federal Centers for Disease Control and Prevention released an analysis of asthma deaths citing a previous study that indicated no evidence to support "the role of outdoor pollution levels as the primary factor driving" the asthma increase.[11] (Ironically, in that same issue the CDC featured a notice from the ALA on asthma prevention, citing smog as a cause.)

Even Harvard's Douglas Dockery, whose epidemiological work the EPA has so heavily relied on in proposing its new standards, admitted in a coauthored 1996 medical journal that "there appears to be no evidence that the prevalence of asthma or asthmatic symptoms in children is associated with chronic exposure to particulate, sulfur oxide, or ozone air pollution." While the authors said that this lack of evidence did not rule out the possibility that high levels of pollution might trigger or worsen an asthmatic attack, air pollution, they said, "does not appear to contribute to the increased prevalence of new cases of asthma, as is often claimed in the popular press."[12]

Something else not often heard in the popular press and never at all from the environmentalists is that the increase in asthma is almost entirely race related. For white children and young adults, there has been essentially no increase. The increase is among blacks,[13] to a point where blacks between the ages of fifteen and twenty-four now have six times the asthma death rate of whites the same age.[14] Although evidence suggests that blacks are more likely to live downwind of factories than whites, utility plant and car exhaust are spread evenly. Is air pollution bigoted? Or is the increase in asthma related to life-style or housing?

In May 1997, researchers reported that the major cause

41

of asthma in inner cities (essentially among blacks) is neither cars nor corporations nor chemical companies, but cockroaches—that insect we all love to hate. It appears that a quarter of all asthma in these areas (which have twice the asthma rate as non–inner cities) comes from those nasty insects.[15]

The asthma is an allergic reaction to the roaches' saliva, decaying body parts, and what environmental regulators might call "tailpipe emissions." Dust mites also pose a serious problem. While completely eliminating roaches can be a daunting task in inner-city apartment complexes and housing projects, just reducing their numbers can provide blessed relief for asthmatic children.

"It's a cruel hoax to lead parents to believe their children will be protected from having asthma if only the EPA clamps down on outdoor air pollution," says Robert Phalen.[16] But we are unlikely to witness a government or activist group campaign against cockroaches, however, when groups like the Chemical Manufacturers' Association make more attractive, and clearly larger, targets.

The Government's SIDS Fibs

"If Congress wants to put up dead babies versus polluters' profits, they should just go ahead," taunted the executive director of the far-left Physicians for Social Responsibility in response to a recent EPA study linking air pollution to Sudden Infant Death Syndrome (SIDS).[17] Yet the apolitical Sudden Infant Death Syndrome Alliance announced that a "causal relationship between particulate air pollution and SIDS has not been demonstrated by the current study."[18]

What's going on here? Yet another effort to invoke the new pollution standards in the name of children. Hence the remarkable "good fortune" of this study appearing in June of 1997, just weeks before the EPA announced what its final proposals would be. It led to such headlines as "Study: Pollution Kills Many Infants" (*Salt Lake Tribune*)[19]; "Air Pollution Puts Babies Susceptible to SIDS at Risk"

(Gannett News Service)[20]; and "SIDS, Air Pollution Linked" (United Press International).[21]

The report found that "infants born in cities with high levels of soot in the air were as much as 26 percent more likely to die of Sudden Infant Death Syndrome (SIDS) and 40 percent more likely to die of respiratory diseases."[22]

But there is a lot more to the story. Galleys of the article were leaked by its lead author, Tracey Woodruff, to radical environmental groups like Clean Air Trust and to selected, sympathetic reporters.[23] Those reporters then eagerly relayed the words of a Trust official: "The science suggests strongly that air pollution is killing infants."[24] It was more than a month before critics and independent-minded reporters could actually read the study in the actual goverment medical journal, *Environmental Health Perspectives.* The need for the spin doctoring becomes apparent when you actually look at the study. It found that infant mortality in the most polluted areas was merely 10 percent higher than in the least polluted areas.

There were other major apparent problems. Curiously, for example, infants in the low birthweight category, who would presumably be more likely to be harmed by pollutants, showed no statistically significant increase in risk. Despite its flaws, EPA administrator Carol Browner seized on the study to push for the new air pollution standards. But the study did not even measure PM2.5, but rather just PM10—which is already being regulated!

But the main cause of SIDS could hardly be outdoor pollution of any type. Remember that the average U.S. *adult* spends more than 90 percent of his life indoors. The percentage of time would surely be greater for the babies who died of SIDS, who were all less than a year old.

The greater problem may be indoor air pollution.

"That study is probably the strongest evidence yet that something other than particulates is the cause of mortality [death], because this is a population that is not going to get much exposure to outdoor air," says Consad's Fred

FIGURE 6–1
ASSOCIATIONS BETWEEN VARIOUS RISK FACTORS
AND SIDS DEATHS
(percent)

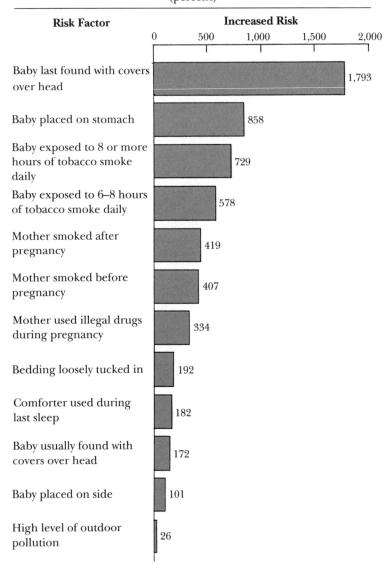

Risk Factor	Increased Risk
Baby last found with covers over head	1,793
Baby placed on stomach	858
Baby exposed to 8 or more hours of tobacco smoke daily	729
Baby exposed to 6–8 hours of tobacco smoke daily	578
Mother smoked after pregnancy	419
Mother smoked before pregnancy	407
Mother used illegal drugs during pregnancy	334
Bedding loosely tucked in	192
Comforter used during last sleep	182
Baby usually found with covers over head	172
Baby placed on side	101
High level of outdoor pollution	26

SOURCE: *British Medical Journal,* July 27, 1996.

Rueter. "I would search for the culprits in indoor air."[25]

What may be even worse than needlessly worrying parents about the causes of SIDS deaths is that it distracts attention from real causes or suspected causes. Thus, while parents are now panicked that outdoor air pollution is associated with a 26 percent increase in SIDS deaths, how many know that a study in the *British Medical Journal* found numerous controllable factors that absolutely swamp that 26 percent? (See figure 6–1.) For example, laying a baby down on its stomach increases its chance of dying by almost 900 percent. Loosely tucking the baby in is an almost four times higher risk factor than is outdoor air pollution. Supporting the indoor air pollution theory is the finding that if the infant is exposed to tobacco smoke for eight or more hours a day, it has a stunning 700 percent higher chance of SIDS! And here's the EPA spreading fear over something associated with a 26 percent higher risk.

Thus, there are two tragedies in the way the SIDS–air pollution study was exploited. The first is having used it to foist terrifically onerous regulations on us that will save no children's lives. The second is that parents are being misinformed about what they can do to protect their kids from SIDS. Any way you look at it, it's pretty sick to play politics with babies' lives.

7

CASAC Capers

Among the evidence that it is on the side of science, the EPA proudly announced that its proposal had been peer reviewed by the Clean Air Scientific Advisory Committee, "made up of nationally recognized experts in many disciplines—physicians, toxicologists, epidemiologists, atmospheric scientists—from academia, industry, and the states." Said Carol Browner in congressional testimony, "Panel members reviewed thousands of pages of materials prepared by the EPA and integrated the best available science."[1] Her testimony lacked the least hint that maybe, just maybe, CASAC did not provide a ringing endorsement of the EPA's decision. Still, to hear the *Washington Post* tell it, "the need for tighter restrictions" on particulate matter was "endorsed overwhelmingly by the EPA's independent scientific advisory committee."[2] The *Boston Globe* told readers "19 of 21 scientists brought in by EPA to review air pollution standards agreed that limits on fine particles are needed." Browner uses the same wording incessantly at congressional hearings.[3]

What CASAC actually concluded on particulates is this: Of the twenty-one members on the panel, only six agreed with both the twenty-four-hour and the annual standards

for PM2.5. Ten disagreed with one or both of the standards. Only four members said the allowable level of PM2.5 should be set as low as the EPA wanted. The rest said the EPA should either set a less restrictive standard, or they did not answer the question at all. Yet these diverse conclusions became an "overwhelming approval" or an agreement of "19 out of 21" people.

Furthermore, according to Roger McClellan, apparently at least several CASAC members voted for some PM2.5 standard not so much because they believed fine particles to be necessarily harmful but because they saw regulating particulate matter as the best way of collecting more information on it.[4] "I feel embarrassed that I'm up here recommending a new PM2.5 standard for getting data, but that is what is compelling me to do that,"[5] CASAC Chairman George Wolff said at a CASAC meeting in May 1996; later he concluded, though, that "there does not appear to be any compelling reason to set a restrictive PM2.5 NAAQS [National Ambient Air Quality Standards] at this time."[6]

Finally, CASAC in its closure letter to Browner in June 1996 stated that the EPA's "deadlines did not allow adequate time to analyze, integrate, interpret, and debate the available data."[7] Such language does not imply overwhelming approval either.

8

Holes in the Ozone Claims

Ozone is something of a two-edged sword. High up in the stratosphere, it partly shields us from harmful ultraviolet light that causes skin cancer and made Brigitte Bardot's skin look like a worn-out saddle. But at ground level, it can make breathing more difficult. And while some have suggested that we simply "ship" ground-level ozone to the stratosphere, thereby solving both the problems of smog and the alleged thinning of the ozone layer, it cannot be done.

What we *can* do is reduce ozone levels here on Earth, but it is not easy. The difficulty arises partly because ozone is not emitted directly in the sense, say, that lead or carbon monoxide is emitted. Ozone forms when two precursor gases, nitrogen oxide and volatile organic compounds, mix together and are heated by sunlight. Without all three ingredients, ozone does not exist. We can do little about the sunshine, which is one reason Los Angeles has both high smog levels and *Baywatch*. (The other natural forces behind Los Angeles's frequent smog are air inversions and the San Bernardino mountains, which trap the pollution.)

Wisely avoiding research on ways to block out or destroy the sun, scientists have instead sought to reduce emissions of the precursor pollutants. In most areas, vehicles are a major source of both VOCs and nitrogen oxide. Hence, we see the third whammy on Los Angeles, the city where the car is king. Power plants produce a large amount of nitrogen oxide as well, and some is produced naturally by vegetation.

The current standard for ozone is no more than 0.12 parts per million (ppm) during the highest hour each day. An area is allowed to exceed this level only three times during the course of three years. In its new regulation, the EPA suggests a range but recommends a maximum of 0.08 ppm averaged over eight hours. It also replaces the standard for excess ppm with a new one that would take the third highest reading from each monitor for each year and average them over three years. Using this number, the EPA says, will be fairer because an area is less likely to be tipped out of compliance because of unusual weather. The Clean Air Scientific Advisory Committee and most experts with whom I spoke agree with the revised procedure for determining excess ppm: score one for the EPA. But tightening the standard raises the stakes considerably.

The new standard is much tougher, so much so that, according to the agency, it would more than double the counties out of compliance, from 105 to 228.[1] One may well suspect the EPA of low-balling this estimate. According to a study by Alan Krupnick and Deirdre Farrell at the Washington-based science think tank Resources for the Future, using a fairly similar standard—an eight-hour measurement at 0.08 with one day of excess ppm allowed per year—would almost *quadruple* the number of noncomplying counties.[2]

Meanwhile, the American Lung Association, along with the American Public Health Association (APHA)—a group that has become increasingly involved in nonscientific, politically oriented issues like handguns and spousal abuse[3]—is calling for an even tighter standard of 0.07 (70 parts per billion), with one day of excess allowed. Adapting this proposal would almost *quintuple* the number of counties in noncompliance: of the 581 counties with monitors, 514 would be violating the law.[4] Yet even the EPA's 0.8 standard would mean "all but ending the hopes of many non-complying areas to attain the standard," according to Krupnick and Farrell.[5]

What would all this regulation accomplish? In recent years, opponents of what they consider excessive environmental regulation have effectively used the slogan, "How clean is clean?" Implicit is the idea that somebody out there wants to reduce man-made pollution levels all the way to zero. There may be no better rallying point for such people than the EPA's new ozone regulation.

First, at least in some areas, tightening the regulations would require eliminating nearly all man-made ozone precursors, and maybe even some natural ones. A large amount of VOC emissions also comes from green plants. Although the media, the EPA, and environmentalists ridiculed President Reagan for saying that trees pollute, a 1989 report by Canada's version of the EPA, Environment Canada, found that 86 percent of the VOCs in that country's air comes from plants, especially trees.[6] In a 1991 report, the National Research Council estimated that natural sources of VOC emissions, resulting in background levels of ozone, are of "comparable magnitude" to man-made VOC emissions.[7] It further states that "this VOC background should be able to generate ozone concentrations that exceed the NAAQS of 0.12 ppm."[8] That number is the *current* standard, not the much tighter EPA standard put into effect recently. In addition, plants put out some of the other ozone precursor, nitrogen oxide.

The Suffocating Cost of Ozone Regulations

What will we get for such reductions? Ozone can cause a range of respiratory problems, including coughing, sore throat, shortness of breath, chest pain, and greater susceptibility to respiratory infections. Those with these or other underlying problems can experience severe effects and may sometimes even require hospitalization. Such effects are usually temporary and indeed may always be temporary. While the EPA's staff paper cites a "possibility" that repeated lung inflammation from ozone can cause permanent harm over a period of decades, the paper concedes that there is little evidence.[9]

Despite ALA claims that ozone at current concentrations actually kills people, even the EPA demurs, saying that "the strength of any such association remains unclear at this time."[10] Of the ten mortality studies cited in the EPA's Criteria Document—including some by Joel Schwartz and Douglas Dockery—only one shows a statistically significant connection between ozone and death. "I think it's probably true to say ozone harms lung function," says Daniel Menzel of the University of California, Irvine, "but to say it's causing mortality is stretching the data."[11]

Indeed, because ozone at the ground level blocks ultraviolet radiation (specifically, UV-B) just as it does in the stratosphere, it is possible that lower ozone levels will actually cost lives and otherwise harm human health. Thinner ozone is thinner ozone; it always allows in more UV-B, causing more skin and eye damage. At least as early as 1995, the Department of Energy alerted the EPA to this fact, specifically regarding the setting of ground-level ozone standards. In an official letter attached to copies of a number of studies, the Energy Department estimated that a mere half-percent decline in ozone would cause twenty-five to fifty new melanoma-related fatalities per year, 130 to 260 cases of cutaneous melanoma, and 2,000 to 11,000 new cases of nonmelanoma skin cancer. Beyond that, there would be 13,000 to 28,000 new cases of cataracts yearly.[12] The DOE is not alone in this concern. Randall Lutter and Christopher Wolz of the Office of Management and Budget, writing in the March 1997 issue of *Environmental Science and Technology/News,* noted that while nobody knows exactly how much ground-level ozone would be released under the new standards, a range of 3,000 to 10,000 increased nonmelanoma cases seemed plausible. Using data from the EPA itself, originally applied to atmospheric ozone depletion, Lutter and Wolz found 37 to 130 deaths per year just from nonmelanoma skin cancer.[13] The EPA is blind to such "collateral damage," to use the military euphemism for friendly forces and civilians killed during bombing raids. Rather than challenge the estimates, it ignores them, since the Clean Air Act does not require the department to take such effects into account. Ironically, given that ozone at U.S. levels is probably never fatal, the EPA proposal itself could well be a killer. The potential for such results seems not to concern the environmentalist groups, who are mightily concerned about ozone thinning up in the stratosphere when industry, by great coincidence, is the culprit.

At what level does ozone become unhealthy for lungs? For most people, that point is above the current EPA standard. But a small portion of the population experiences significant loss of lung function at relatively low ozone concentrations; these people are called "responders."

Unless exercising heavily, however, even the most sensitive individuals do not normally experience significant reduction in lung function after one to three hours of exposure at the current 0.12 ppm level. A 1990 study in the American Lung Association's *American Review of Respiratory Disease* concluded that while "some individuals experienced substantial pulmonary distress" at low concentrations, "others will not experience any from such exposures."[14] Based on the EPA's estimate of the relationship between changes in forced expiration and various combinations of exercise levels and ozone concentrations, typical subjects experience less than a 10 percent loss in lung function at ozone levels more than four times the current standard. Even during very heavy exercise, lung function is usually reduced by less than 10 percent at 0.24 ppm—twice the current standard, and almost three times the proposed standard.[15]

In a 1987 review of the ozone standard, the Clean Air

Scientific Advisory Committee suggested that such mild responses to ozone not be considered an "adverse" respiratory health effect. The committee concluded that these mild physiological changes would probably not be considered medically significant and should not interfere with the normal activity of most individuals. Indeed, mild decreases in pulmonary function may go unnoticed by the typical person, given the lungs' substantial reserve capacity.[16]

Furthermore, the studies in which persons are made to exercise are necessarily artificial in that subjects are encouraged to keep going unless they felt they were experiencing great discomfort. In the real world, mild discomfort would probably cause all but the most determined people to reduce their exercise.[17] Finally, evidence suggests that our lungs adapt to ozone exposure in somewhat the same way they adapt when we move from sea level to the mountains and stay a few days.[18] This is not to say, though, that the adaptation is complete.

How Low Can You Go?

For the most susceptible among us, the responders, however, ozone may cause problems even at "background-level." Those are the problems that would occur even if *Homo sapiens* had never crawled out of the proverbial primordial swamp and built cars, factories, and Lawn Boy riding mowers. In other words, no matter where the standard is set, some responders, especially those exercising heavily, will be made uncomfortable or will even experience pain. As the 1996 ozone CASAC panel put it, there is no "bright line," allowing us to fix a level of safety. It also pointed out that of the various standards the EPA has proposed, none is "significantly more protective of public health than its alternative proposal."[19] CASAC labeled the choice of an ozone standard a "policy judgment,"[20] and as CASAC Chairman George Wolff notes, "although the panel members' opinions differed, none supported the lower end of EPA

53

staff's recommendations (essentially the ALA recommendation), and the majority of the members stated a position which included . . . equivalent stringency to the *present*" standard (emphasis added).[21]

Fortunately, there seem to be few such responders to ozone in the general population. The EPA staff paper estimates that from one to three summertime respiratory hospital admissions per day may result from each 100 parts per billion of ozone per million exposed persons. In other words, in a city of 1 million three added summertime admissions per day might occur if the ozone level tripled from a background concentration of around 0.05 ppm to 0.15 ppm.[22]

Similarly, making the assumption that ozone can cause asthma attacks, the EPA staff paper projects that the difference between hospital admissions for asthma in the New York City area under the proposed standard would be a total of ninety.[23] This number is just six-tenths of 1 percent of the total of about 15,000 asthmatic hospital admissions in the New York City area over the ozone season.[24]

At the February congressional hearings, Sen. John Chafee (R-R.I.), widely considered one of the GOP's most ardent environmentalists, expressed dismay at this figure. "Am I missing something here," he asked New York University's George Thurston, "or are we dealing with very minor health improvements here?" Thurston replied, "I guess it comes down to whether you're one of those people."[25]

In a sense, both are right. If a disease afflicted just two Americans and you or—especially—your child happened to be one of them, you might well think it worth an expenditure of billions of dollars a year to prevent it. But lawmakers do not have the luxury of thinking that way. Because they are spending the money of everyone they represent, they should make laws for the good of all. I know what it is like to suffer with asthma; I had it as a child, as did one of my brothers. Sometimes I felt as if Baby Dumbo were sitting on my chest. But lawmakers must consider whether individuals can take action themselves to reduce or eliminate problems, without relying on government interdiction and the expenditure of other people's money. Fortunately, people these days have access to far more information about what causes asthma and to effective new medicines that were not available even a decade ago—most important, inhalable steroids that have been shown to be twice as powerful as the previous generation of drugs.[26] Yet, points out Dr. Henry Milgrom of National Jewish, a hospital in Denver, "only about 50 percent of inhaled [asthma] medication is taken as prescribed."[27] Even if ozone were causing asthma, a minuscule increase in the medicine usage rate would readily outweigh anything we could hope to gain with massive spending on reducing this pollutant.

We need to ask if there are not better uses of such spending than, at best, to make a slight dent in preventing attacks of a disease that is becoming ever more treatable and preventable on a personal basis. A better approach might be more research into learning what really causes asthma and then developing better ways to prevent and treat it. Yet this year the National Institutes of Health (NIH) will spend just $86 million on asthma research,[28] while estimates on the cost of fully complying with the proposed new ozone standard, as we shall see shortly, run into the tens and even hundreds of *billions* of dollars. Another approach would be to educate parents more on what they can do to mitigate the suffering of their children. How many inner-city parents are aware that the best thing they can do to help their children breathe better is to reduce their home's roach problem?

With air pollution, as with so many other political issues, people often point to Europe as an example of what we should be doing here. The comparison is commonly false—so too with pollution. Thus, Morton Lippmann in his Senate testimony noted that the air quality guideline for ozone of the World Health Organization–European region, adapted in 1996, is actually considerably below even

EPA's proposals.[29] Implicit in this guideline is the notion that the Europeans have decided that such a pollution level is both achievable and necessary. Who are we (or, more specifically, American industry) to say American lungs do not deserve as good treatment as Dutch or Danish lungs?

But the key word here is *guideline,* instead of *standard.* The European rule contains no enforcement power, no "teeth." The individual European governments made sure of that. Thus Athens continues to have air that makes the air in Los Angeles seem absolutely pristine. Even the collecting of pollution data in many European countries relies on practically Stone Age techniques compared with ours.[30] The WHO guidelines do nothing more than say a low level of ozone would be "a good thing." Probably not many Americans would desire such "protection." Lippmann, to his credit, did not invoke the extremely stringent pollution guidelines in Russia, which also happens to be one of Eurasia's most polluted countries.

Some American authorities, such as the Chemical Industry Institute of Technology's Roger McClellan, believe a combination of standards and guidelines could be useful here. "I think we ought to be giving consideration to that as we make further progress in reducing air pollution," he says. "As we now approach such low levels, I believe it's appropriate to give states and municipalities a greater degree of latitude with how they deal with air quality issues in their particular area." He invokes the backwardness of a "one size fits all" attitude.[31] He notes that allowing for such variations is particularly relevant given the widely different constituents of PM2.5 from area to area, as the EPA's data in Washington, D.C., Phoenix, and the San Joaquin Valley clearly show. In the District of Columbia, for example, over 46 percent comes from sulfates emitted by oil- and coal-fired utilities, institutional boilers, and smaller combustion sources. Only 5 percent comes from soil, and all that is from dust kicked up from paved roads. In contrast, in the San Joaquin Valley, only 11 percent comes from sulfates, while soil constitutes over 7 percent and almost all that is from agricultural tilling and erosion. In Phoenix, soil contributes fully 16 percent of total PM2.5, with the majority of that coming from construction. In San Joaquin, we see that wood burning from residences, managed burns, wildfires, and structural fires contributes far more to the PM2.5 level than do sulfates.[32]

Except for those who live in one of these three areas, nobody can predict exactly what a PM2.5 regulation will mean for residents—who will be hit hardest and who will be hit least. But already the proposed regulation has heightened anxiety among the nation's farmers, who are still struggling to meet the current PM10 standard. The American Farm Bureau and twenty-four other agribusiness groups have expressed their concern in a letter to the EPA,[33] and a Farm Bureau representative testified before Congress in April 1996 that a PM2.5 regulation has the potential to "put American agriculture out of competition with other countries and put agricultural producers out of work," because farmers in other countries will bear no such burdens.[34] Says Wayne Coats, of the Office of Arid Land Studies at the University of Arizona, the new EPA standard "would significantly affect agriculture" because it "likely will be increasingly blamed as a source of particulates." He believed the standard would result in "many restrictions [being placed on] agricultural field operations."[35]

Unfortunately, notes McClellan, taking into account different geographic areas and the different constituents of particulate matter is not allowed under the Clean Air Act.

9

To Enforce the Impossible Dream

When the original Clean Air Act was enacted in 1970, it was based on the assumption that every pollutant has a threshold below which the concentration has no ill effects. It could well be argued that the proposed EPA ozone standard is a move in that direction, but the direction is toward a goal that may be impossible to meet.

The Clean Air Scientific Advisory Committee said just this in its final report to Carol Browner, stating that "the paradigm of selecting a standard at the lowest-observable-effects level and then providing an 'adequate margin of safety' is no longer possible."[1] The EPA itself appears to have tacitly admitted this impossibility in coming up with a cost-benefit analysis for both ozone and PM2.5 that assumes incomplete compliance. The problem therefore lies with the wording of the Clean Air Act. Congress should revise the act's goal to "protect the public against *unreasonable* risk of *important* adverse health effects," recommend Stephen Huebner and Kenneth Chilton of the Center for the Study of American Business at Washington University in St. Louis[2] (emphasis in original). Whatever the wording, a quarter-century after its enactment, the Clean Air Act has clearly run head-first into brutal reality and been shown to be sorely outdated.

The EPA's Plan to Stimulate the Economy

While the Clean Air Act forbids the EPA to take economic costs into account in *setting* health standards, an executive order of President Clinton requires it to make a cost-benefit analysis anyway.[3]

Cost-benefit analysis means seeing what benefits a given regulation will yield compared with what it will cost to implement the regulation. The idea is that while it is often easy to see the benefits of any given new rule, any possible financial downside should usually be taken into account as well. In other words, we cannot look at just one side of the equation. Originally, this tool was a remarkable weapon in the battleground of ideas. It caught the more strident regulators and their activist allies, such as the environmental groups, completely off guard.

But regulators soon developed a shield against cost-benefit analyses, the kind of shield that could also be wielded as a weapon. All they had to do was to find that the regulations, if looked at the right way, if tweaked and pinched, if held up to just the right light, could *always* more than pay for themselves.

So it goes with the EPA's cost-benefit analysis of the tightened standard for particulate matter. It estimates that the annual costs of partial attainment would be $6.3 billion and that benefits would amount to $58 billion to $119 billion.[4] (This number would have to be revised downward somewhat in light of the decrease in the lives-saved estimate from 20,000 to 15,000.) Specifically, the EPA claims that each life saved is worth $4.8 million,[5] to which the costs of the illnesses and damage to trees and buildings that would also be prevented would need to be added. There we have it: the cleaner we make the air, the more we boost the economy. Who could argue with that?

Lots of people, apparently. One reason for skepticism is that we have seen this kind of EPA clean air hocus-pocus before. In 1990, researchers Michael Hazilla and Raymond

Kopp of Resources for the Future determined that environmental policies had reduced the gross national product almost 6 percent by that year. They calculated that this percentage translated into over a million lost jobs.[6] Of course, there are benefits other than financial ones from environmental regulations. But a better environment is something we have to buy, like a better house or a better car. If we want a better house, we must be prepared to pay for it. The EPA could have pointed this out.

But the EPA did not choose that route. It put its own calculators into overdrive and came up with the amazing "discovery" that the total benefits from what is generally thought of as the most onerous of the environmental regulations, those pertaining to clean air, had by 1990 actually reaped the nation a $1.3 trillion windfall! That amounts to almost a quarter of the entire gross domestic product.[7] Christ's miracle with the fishes and the loaves pales beside what the EPA claims to have accomplished with its clean air regulations.

Still, there is nothing implicitly wrong with the assertion that environmental regulations can actually save money. The "boy who cried wolf" was eventually correct; maybe this time the EPA is, too. Let's look first at the benefit side, then the cost side.

To understand benefits, we have to assign some sort of value to human life. This valuation is obviously not subjective. If it were, safe to say, most of us would assign huge—if not infinite—numbers to our parents, children, and friends, while people like Stalin, Timothy McVeigh, or anybody who pops his gum in your ear would be deep in the negative dollar category. Allocating a value to life, though, is an objective way of allocating risk. We have only so many dollars to spend for risk reduction, and we need some sort of yardstick for what a life is worth to be able to assign those dollars.

With that in mind, consider the EPA's assignment of $4.8 million for each life saved from premature PM death.[8]

The Department of Transportation (DOT) uses a considerably lower figure, $2.7 million as of 1996.[9] That disparity is interesting, but it gets more interesting yet. The DOT is inherently concerned with accidental deaths, which can happen at any age, but for its calculation it assumes the death at age forty of a person in good health.[10]

In stark contrast, the particulate studies that find any premature dying at all tend to find it among "vulnerable individuals, primarily the elderly and individuals with pre-existing respiratory disease," as the EPA staff paper on particulates puts it. Thus, the paper continues, "some of the mortality associated with short-term pollution is occurring in the weakest individuals who might have died within days even without PM exposure," something it calls the "harvesting effect."[11] Numerous researchers told me they believe that the epidemiological studies probably did show harvesting, and Johns Hopkins's Jonathan Samet has also suggested this as a possibility.[12]

In any case, even the Natural Resources Defense Council admits, "The elderly and those with heart and lung disease are at greatest risk of premature mortality due to particulate air pollution."[13] Browner herself told Congress that the average shortening of life was from one to two years.[14] Thus it is a bit disingenuous for groups like the NRDC [15] and newspapers like the *Los Angeles Times* [16] to compare deaths from particulates with deaths from vehicle accidents or from AIDS, which takes the lion's share of its victims from the 30–39 year-old age range.[17]

Be it two weeks or two years, either hardly justifies valuing the slight extension of a very old or a very sick person's life at millions of dollars. Indeed, a recent study found that people from Sweden would value their own lives, after the age of seventy-five, at only $400 to $1,500 per year.[18] Even assuming that the proposed air pollution rules would extend life by two years, what the Swedes peg at $800 to $3,000, the EPA puts at $4.8 million. Sweden, not incidentally, is one of the most health- and safety-conscious countries in the world, with stringent environmental laws and an auto-mobile fatality rate less than half that of the United States.[19] When the EPA says the value of extending a person's lifespan is equivalent to the cost of Arnold Schwarzenegger's home and the Swedes say it is equivalent to a 1981 Volvo, there is something wrong with this picture.

10

Spinning Out on the Cost-Benefit Curve

What about the cost side of the PM2.5 standard? The EPA puts the price at about $6.3 billion a year, an amount in addition to that already being spent for PM10 reduction. While that is not exactly pocket change, PM2.5 would hardly be the most onerous regulation in this country. The $6.3 billion, though, is for "partial attainment." In all environmental regulations, each new bit of cleanliness costs more than the bit before. Economists illustrate this effect with an analogy to picking fruit. At first the picker can just reach up and grab the apples from low branches. But then he finds he needs ladders, and then taller and shakier ladders. Food metaphors aside, maybe reducing the first ton of emissions costs just $50 per day, but the tenth ton is $200 and the twentieth is $500. Full attainment does not cost just twice as much as half attainment: ultimately, it could cost hundreds of times more.

As cities try harder and harder to comply fully with PM regulations, costs skyrocket. The closer they get to full attainment, the more dear it becomes. How dear? The EPA estimates that full attainment would require removing an additional 13 micrograms of PM2.5 per cubic meter of air nationally, at an annual cost of far more than $1 billion per microgram removed.[1] The agency does not say how much over a billion, but one EPA-commissioned study using data from Philadelphia estimated it at over $4 billion. Even that did not allow full attainment.[2] Using the Philadelphia numbers, the Center for the Study of American Business estimated that the new PM2.5 regulations would cost the nation $55 billion or more each year.[3] The Reason Public Policy Institute, in a report released in June 1997, put a range on it—somewhere between $70 billion and $150 billion a year.[4]

Bronchitis for Sale—Half a Million Bucks

What about the EPA's cost-benefit analysis for ozone? Oddly, the agency has established a cost-benefit analysis for ozone reduction but provides two estimates, and neither is exactly for the tougher standard. One applies to a somewhat stricter standard than the one promulgated, and the other to a somewhat looser standard. The tougher one shows partial attainment costs of $2.5 billion per year with benefits of $100 million to $1.5 billion. The looser one has costs of $600 million per year and benefits ranging from nothing to $500 million. Thus the EPA does not even pretend that the ozone regulations will pay for themselves.

This result, however, does not come from the agency's being seized by a fit of honesty. According to other calculations, even the EPA's admission that the ozone regulations are no free lunch grossly understates the problem. Again, on the benefit side the agency makes assumptions that defy credibility. Among them is that preventing a case of nonfatal bronchitis is worth a stunning $587,500.[5] When I was a child, I developed bronchitis like clockwork every single month for years. While it was not fun, I would gladly have suffered the bronchitis for a lot less than half a million a case. (Safe to say, I still would!) The Treasury Department's Office of Economic Policy, in a December 1996 memo to the EPA, also expressed incredulity at this figure.[6]

As for the costs, once again the EPA keeps expenses low in its model by simply assuming that regions in nonattainment will at some point just give up and suffer Washington's punishment. But what if the areas keep trying? What if they keep escalating spending to comply with the ozone standards?

This is exactly the problem Alicia Munnell, a member of the President's Council of Economic Advisers (CEA), considered in a December 1996 memo to the EPA, a memo that did not come to public attention until three months after it was sent. In it, Munnell said that the EPA's use of only a partial cost analysis "understates the true costs of stricter standards by orders of magnitude."[7] How many orders? According to her, "CEA estimates indicate that the cost of full attainment could be up to $60 billion,"[8] which would show net costs of almost $60 billion a year since the ozone standard would provide virtually no benefits.[9] By comparison, the entire federal budget for medical research is about $15 billion.[10]

Using data from three different cities—Chicago, Fresno, and Philadelphia—the CEA challenged the EPA's assertion that additional ozone-causing pollutants can be removed at a cost of $3,000 to $10,000 a ton, asserting that the actual cost would probably be $30,000 to $80,000 a ton.[11]

Other analyses have also shown that the price of meeting the new ozone standard would be staggering. A report from the Reason Institute estimates full attainment at between $20 billion and $60 billion a year,[12] while George Mason University's Center for the Study of Public Choice puts it somewhere between $54 billion and $328 billion.[13] Yet we have no reason to think that even an 0.07 standard—much less the EPA's new 0.08—will ultimately satisfy the regulators. Indeed, Barry Levy of the American Public Health Association told Congress that "levels of ozone at 0.07 ppm still cause measurable health problems that the public health community cannot accept," except perhaps right now for political expedience.[14]

If the purpose of the new regulations is truly to save lives, there is something quite perverse about the effort. Economists have long been piling up mountains of evidence showing that, whether comparing groups within a country or comparing nation with nation, those with more money to spend live longer. The more money you have, the better medical care you can afford, the fewer risks you have to take, and the more you can spend on safety devices or simply safer things. "Wealth equals health," goes the saying. Or to put it in negative terms, "Anything that retards economic growth also retards environmental cleanup and consigns millions to squalid and untimely deaths." Thus says Indur Goklany, Interior Department manager of science and engineering in the Office of Policy Analysis.[15] In the study cited in chapter 4, lack of income corresponded to a much higher rate of respiratory problems and heart disease. Using data derived from self-reported value of life surveys, Harvard University professor W. Kip Viscusi has calculated that generally speaking "every $50 million spent on regulation induces one statistical death."[16] Using a different model, one based on household disposable income and deaths from all causes, the Reason Institute comes up with a much higher figure—specifically that every $45 million spent in compliance would cost one statistical life.[17] (See table 10–1.)

Obviously, some of these ranges are quite large. But adding together ozone and PM2.5 and averaging the lowest estimates with the highest provides a figure of about 52,000 deaths a year from the EPA's new standards—far more Americans than die in vehicular accidents.

Mind you, it is not Washington bureaucrats or environmental lobbyists who will foot the butcher's bill. Of those who die, somewhat more than half will be those earning less than $15,000 annually and slightly fewer than a quarter will be black, according to the Reason Institute. We won't know the victims' names. Their relatives won't be interviewed by CNN or CBS. But they will perish nonetheless.

TABLE 10–1
ESTIMATED COSTS OF THE EPA'S NEW AIR REGULATIONS IN
DOLLARS AND LIVES

Estimator	Estimated Cost of Regulations ($ billions)	Estimate of Lives Lost Viscusi formula	Estimate of Lives Lost Reason Institute formula
Ozone Regulations			
Council of Economic Advisors	60	1,200	13,000
Reason Public Policy Institute	20–60	400–1,200	4,500–13,500
George Mason Center for Public Choice	54–328	1,080–6,560	12,000–73,000
Fine Particles Regulations			
Center for Study of American Business	55+	1,100	12,200
Reason Public Policy Institute	70–150	1,400–3,000	15,500–33,300

SOURCES: Viscusi formula estimate of lives lost data derive from W. Kip Viscusi, "The Dangers of Unbounded Commitments to Regulate Risk," in Robert Hahn, ed., *Risks, Costs, and Lives Saved: Getting Better Results from Regulation* (New York: Oxford University Press, 1996), p. 162. Reason Institute formula estimate of lives lost data derive from Ralph L. Keeney and Kenneth Green, *Estimating Fatalities Induced by Economic Impacts of EPA's Proposed Ozone and Particulate Standards* (Los Angeles: Reason Public Policy Institute, June 1997), pp. 7–9. CEA data derive from Munnell, draft memorandum to Art Fraas, p. 2. Reason Public Policy Institute data derive from Smith et al., *Costs, Economic Impacts, and Benefits,* pp. 16 and 26. George Mason Center estimates derive from Susan E. Dudley, *Comments on the U.S. Environmental Protection Agency's Proposed National Ambient Air Quality Standard for Ozone* (Fairfax, Va.: George Mason University, Center for the Study of Public Choice, March 12, 1997), p. 8 and appendix A, p. 42. Center for Study of American Business data derive from Hopkins, "Can New Air Standards," p. 16.

Where Has All the Low Fruit Gone?

The EPA-environmentalist response to this argument is, essentially, "Look, we've heard it all before." Industry always claims that new regulations will cost a lot more than they end up costing, in part because industry disguises the fact that Americans are the most ingenious people on earth and can always come up with new, cheaper ways of doing things, including reducing pollution.

And they are right—to a point. Generally speaking, in these matters, when it comes to new regulations, the EPA and the environmentalist groups claim the rules will be a mere pinprick. Industry says the regulations will be so devastating as to bring the U.S. economy to its knees. And when all is said and done, the tighter restrictions end up costing a lot more than the proregulators said and a lot less than what the antiregulators said. Concedes the American Petroleum Institute's (API) Thomas Lareau, "Each interest has its own culture; we tend to look at worst-case on cost, and the environmentalists look at the best case."[18]

But his concession comes with several important notes. First, even the CEA's estimate of the cost of the new regulations makes the EPA's look outrageously, irresponsibly low. These are the government's top economists, who report directly to the president. Second, says Lareau, is that "we have a long history of trying to prevent" these pollutants "and we have found most of the easy, inexpensive ways to prevent pollution."[19] In other words, the low-hanging branches have been stripped clean of fruit.

Indeed, when it comes to spending on ozone or on particulates, we run up against two different and disheartening cost curves. First, each new increment of pollution reduction costs more and more. Second, with each reduction in pollution, we receive fewer health benefits. The first ton of reduction may have sent one less person to the hospital. But soon we had to reduce emissions by ten tons to send one less person to the emergency room. And then it became a thirty-ton reduction to have the same effect. So the more we spend, the less we incrementally reduce emissions. The more reduction in emissions, the less incremental reduction in health problems. As regulations become stricter and stricter, the added benefits realized become fewer and fewer.

But the environmentalists do not accept the low-hanging fruit idea. "There are cost-effective solutions like energy efficiency programs and the use of cleaner fuels such as natural gas and renewables," said David Hawkins, general counsel of the Natural Resources Defense Council, in a press release. But already the vast majority of new power plants being built are fueled with natural gas. Converting a coal-fired plant that serves 50,000 homes to cleaner natural gas will cost about $10 million at the least. But gas does not arrive by train like coal. Unless the plant has the great fortune of sitting on top of a gas reserve, the gas has to be brought in via a pipeline that, depending on the distance from the gas field, can cost another $100 million or so. According to John Schneibel at the Electric Power Research Institute, this could double or triple consumers' electricity costs.[20] Every hot summer some elderly people die of heat stroke because even though they had air conditioners, they felt they could not afford to run them. How many more deaths would there be if electricity costs doubled or tripled?

As for renewables, the only ones that were ever cost efficient were hydroelectric dams and nuclear power. We have nearly run out of places to put dams, and of the few dam sites left, many are opposed by environmentalists.[21] Environmentalists also killed off atomic energy so that no new nuclear plants are being built and operating ones are being shut down prematurely. As for other renewables, as Ralph Cavanagh of the NRDC recently put it, "America gets less electricity from solar, wind, and geothermal today than it did five years ago."[22] Together, these three produce about a fifth of 1 percent of our energy.

Still, say some environmentalists, *Damn the costs and*

full speed ahead! Industries "can talk all they want about costs," one anonymous Sierra Club official told the *Utility Environment Report,* "we're going to crush them."[23] Yes, along with their employees and customers.

To Market, to Market

One weapon the EPA and environmentalists brandish to show that industry's cost projections are grossly pessimistic is that back when both sides were haggling over acid rain legislation, some industry projections were that the price of low-sulfur coal—which greatly reduces the emissions of chemicals that form acid rain—would shoot up to as much as $1,000 per ton. Yet today, as Carol Browner has eagerly noted, emission allowances for low-sulfur coal are being sold for about $100 a ton. We can expect a replay with the new rules as well, she indicated.[24] The acid rain situation, though, was unique. First, there was no emissions trading system at the time, and it is generally agreed that this system did much to reduce emissions costs. Factoring in what does not exist is difficult. Moreover, this system was not a function of technological innovation, so that argument does not support the environmentalist and EPA position that such innovation always comes to the rescue with something easy and cheap. The emissions trading system has also been done and is no longer a new option.

Second, an unforeseen effect of the acid rain legislation was to cause low-sulfur coal suppliers to increase the supply to such a point that the price fell sharply. The claim that the experience with low-sulfur coal and the acid rain legislation can be used to predict what will happen to the cost of stiffer controls on particulates and ozone has no foundation.

But the situation is a lot worse than that. Tighter emission standards could actually hurt or even destroy the entire emissions trading program. Here's why: For trading to work, somebody—say, the Red Power Plant—has to be able to reduce its emissions below the required level. Then somebody else—say, the Blue Power Plant—concludes that it will be cheaper to get Red Power to reduce its emissions by so many tons than for Blue Power to do so. Blue Power, then, pays (by buying Red Power's credits) Red Power to reduce Red Power's emissions beyond what the law requires. The pollution levels go down, Red makes a profit, and Blue spends less money than it would have to if it had to make reductions at its own plant. Everybody is happy.

But "when you require everybody to do everything they can, you've removed that essential bit of slack or choice, or whatever you want to call it that allows the market to be effective," points out Gordon Hester, in the environmental risk analysis program at EPRI. "Right now, it's 'we do it, or we pay somebody else to do above and beyond what they were required.' When you crank down controls to a point where there can be no above and beyond, then you've destroyed the market system."[25]

So the market system that environmentalists initially eyed with great suspicion, but that many of them now eagerly support, could crumble. Short of failure of that system, new, stricter emissions regulations will keep more and more plants from being able to go "above and beyond." The market system will thus be hamstrung.

11

Barbecue Police and Body Odor

The regulators and activists brush aside arguments that the new EPA standards could have a tremendous impact on American lifestyles. The standards are "not about outdoor barbecues and lawn mowers," Carol Browner has repeatedly said in congressional testimony.[1] Smearing industry's claims that these would be affected by calling them "junk science," she said: "These are scare tactics. They are fake. They are wrong. They are manipulative."[2] The Environmental Working Group's (EWG) Richard Wiles says, "No one is proposing to regulate barbecue grills." Talk about barbecues and lawnmowers is "crazed propaganda," says Frank O'Donnell, executive director of the Clean Air Trust.[3]

Yet back in 1994, the EPA had already announced plans to regulate lawn mowers. "The small gasoline engines that Americans use in yard and garden work are a significant source of air pollution,"[4] Browner said at the time. In 1996, the EPA did promulgate emissions standards for lawn mowers, though they were fairly mild as a result of "the antiregulatory climate in Washington and GOP efforts," as the *Christian Science Monitor* put it.[5] A document from the Pentagon noted that to comply with "the current ozone standards, EPA proposed [restricting the use of] even lawnmowers and other small engines."[6] Also under Browner, the EPA has begun the process of regulating powerboats and jet skis.[7] How quickly the administrator forgets such things. Or does she believe that such regulations, while necessary under *current* standards, will somehow become less so under the tighter new, as yet unenforced ones?

Already one state, California, does regulate barbecue grills, along with such other consumer items as leaf blowers and paint. Denver severely restricts the use of woodburning fireplaces and has outlawed the installation of such fireplaces in new homes. Burn a log: go to jail. (Actually, it is a fine.) Regulators in San Francisco have even urged residents to refrain from using aerosol deodorants and alcohol-based perfumes to reduce ozone-creating gases.[8] San Francisco could soon become as well known for body odor as it is for its cable cars and stunning views.

In his effort to assure us that barbecues would not be touched, EWG's Wiles explained, "It would take thousands of barbecues grilling 24 hours a day, every day, to equal the air pollution from one large steel mill."[9] The point he misses is that there are only about 1,100 iron and steel foundries in the country, while according to the Barbecue Industry Association, there are more barbecues in the country than there are households.[10] The number of foundries declines each year, while the number of barbecues grows.

Further, factories and electric utilities are often located outside towns, where the vast majority of particulates reach neither lungs nor pollution monitors. Residential burning, by definition, is right where people live. Those particular particulates will play a bigger role in putting an area into noncompliance, thereby drawing the eye of local regulators.

Jerry Martin, a spokesman for the California Air Resources Board, stated the obvious: "As we have gotten the biggies under control, the smaller sources of pollution become more important."[11] This will mean more mom-and-pop places, such as bakeries and dry cleaners, and it will mean households. It will affect barbecues; it will affect other recreation; it will affect gardening and lawn maintenance. It will affect our personal lives in any number of ways. It absolutely has to; there is no place else to go. "You're not going to get this from digging in the bin of the traditional pollutants," says Roger McClellan. "You're going to have to start looking at new things."[12]

When confronted with some of this, Browner insisted to Congress that blaming the EPA would be completely wrong because these decisions are made by local authorities. This attitude infuriated several congressmen, who remained unimpressed by Browner's distinction between the judge who orders the execution and the man who pulls the switch. In any case, Browner has gone so far as to claim, "I do not believe lifestyle issues will be involved."[13] To pursue the analogy, the last statement says that the switch would not even be pulled—even as it is now being yanked under current pollution standards. Even the administration's own assistant secretary of transportation, Frank Kruesi, expressed fear in a memo that the new pollution standards may "require lifestyle changes by a significant part of the population."[14]

Lawn mowers and barbecues aside, the tightening of regulations on "big business" hardly leaves Mr. and Mrs. Joe Average untouched. All these regulatory costs are passed on to consumers. If a power bill jumps because a utility had to switch from coal to natural gas, if a dry cleaner has to switch to more expensive chemicals, the consumer will pay for it. Conversely, for those pulling down the salary of the head of the EPA or the chief counsel at the National Resources Defense Council, such things would not even be a dent in the household budget.

"One local regulator told me that in some areas he regulates," Robert Phalen says, emissions "would have to be cut to one-tenth the current level to meet the PM2.5 and the ozone standard. How are you going to do that? Driving cut to a tenth, home emissions cut to a tenth, industrial emissions cut to a tenth?"[15]

The EPA's Small Business "Baloney"

Another act of EPA arrogance was the agency's refusal to follow the procedures of the Small Business Regulatory Enforcement Fairness Act. Signed into law in March 1996, SBREFA would require the EPA to convene a small business advocacy review panel, essentially to act as a devil's advocate.[16] Since everybody, left and right, celebrates small businesses, an EPA-convened panel that poked major holes in the agency's conclusions would look very bad indeed for the agency. So the EPA said, in effect, "Panel? We don't need no steenking panel!"[17]

Why not? According to the agency, the standards would not have "a significant economic effect on a substantial number of small entities." In any case, the EPA will not make the regulations, just force others to make them.[18]

But the nation's small business groups seem to think that is nonsense. "That's a cop-out; that's baloney," K. C. Tominovich, manager of legislative affairs of the National Federation of Independent Businesses, commented. "We're not in position to say 500,000 of our 600,000 members will be affected, but we want the EPA to do the analysis."[19] Todd McCracken, president of National Small Business United, said, "It is clear that the EPA has circumvented the provisions of SBREFA by claiming that a health standard does not necessarily impose compliance measures on small entities."[20] Other groups opposing current implementation of the standards include the National Black Chamber of Commerce, the National Indian Business Association, the U.S. Pan Asian American Chamber of Commerce, the U.S. Hispanic Chamber of Commerce, and the National Association of Neighborhoods.[21]

"I represent 24,000 Native American businesses, both individual and tribal," says Peter Homer, president and CEO of the National Indian Business Association. "Most think they're already overregulated. Of our priorities on environmental issues, we have ten or eleven ahead of these, but the EPA doesn't want to mess with them."[22]

Even the Clinton Small Business Administration (SBA) is convinced that the EPA's position is absurd. "We urge the agency to rethink its position," wrote the SBA's chief counsel for advocacy, Jere Glover, to Administrator Browner in a November 1996 memo on the proposed ozone standard. Glover noted that the EPA's claim of no "significant impact . . . would be a startling proposition to the small business community," and that the "EPA's own analysis" showed the new standards "will unquestionably fall on tens of thousands, if not hundreds of thousands of small businesses." Anywhere between ten and fifty-four different types of industries that generally have fewer than one hundred employees "would face costs in excess of 10 percent of sales" because of the new ozone standard alone, Glover said. He then added, in bold, *Thus, this regulation is certainly one of the most expensive regulations, if not the most expensive regulation faced by small businesses in ten or more years."*[23]

Meanwhile, a memo from the office of the secretary of agriculture said, "We share the concerns of the Small Business Administration regarding the potential impacts of these regulations on small businesses. Can EPA address these concerns before the final rule is issued?"[24] It could have, but it did not.

Glover also rejected the claim that the EPA is off the hook because it would not promulgate the regulations itself—rather, state and local authorities would. Glover noted that there is no real distinction between forcing somebody to do something and doing it yourself. Further, he added, some federal regulations already in place will automatically be changed if standards are tightened, thereby making the EPA itself the direct enforcer.[25] No matter. Six months after receiving Glover's memo, Browner used exactly the same copout in testimony before Congress.[26] At the Nuremburg trials, the common excuse was "I was only following orders." Browner has appeared to invert that to "I was only giving orders."

12

Distractions and Double Standards

By and large, the mainstream media seem to have a simple formula as to who is trustworthy in environmental debates and who is not. The EPA and the environmentalist groups want only what is good for us; industry wants only what is good for industry. Industry ties must always be emphasized; environmentalist or EPA ties can be ignored. Consider a January 1997 *Wall Street Journal* article about a gathering of prominent pollution scientists at an Annapolis, Maryland, think tank. It noted that the think tank was funded by the National Association of Manufacturers and that most scientists invited were skeptical of the new EPA standards. The story quoted Joel Schwartz as saying the meeting was an industry "set-up job." Schwartz, meanwhile, was merely identified as a Harvard epidemiologist,[1] without it being noted that until recently worked for the EPA, that in 1996 he received a $196,000 grant from the EPA to study the health effects of pollution on children, and that he has been a crucial supporter of the NRDC in its push for pollution regulations even tougher than the EPA's.

Similarly, when the NRDC released its *Breath-Taking* report, such major media as *U.S. News & World Report,* the *Los Angeles Times,* and the *Washington Post* would not allow critics to get a word in.[2] Headlines, many on front pages, included: "Our Breath-Taking Air," "Microscopic Killers," "Fighting a Microscopic Killer," and "Dirty Air's Death Toll Estimated at 64,000."[3]

This double standard is most pernicious when used to question the trustworthiness of studies. When a study is industry funded, this is always pointed out—along with a powerful hint that because it is industry funded it should be ignored or, at the least, discounted. When a study is funded by the EPA or by an activist group such as the NRDC or the American Lung Association, the study is considered completely impartial and trustworthy.

That attitude is rather like a judge telling a jury to take everything the plaintiff's lawyer says with a grain of salt because, after all, the plaintiff's lawyer is being paid to say what he says—without noting that the same is true of the defendant's lawyer. Furthermore, it is often if not usually the case that the scientists taking industry money for one study took EPA money for an earlier study, and they may well take EPA money for the next one. Sometimes the very same scientists go from being trustworthy to being untrustworthy to being trustworthy. Finally, it is hardly a foregone conclusion that industry-funded studies will find and publish data that serve industry.

Consider the Electric Power Research Institute. EPRI has repeatedly given money to people like Joel Schwartz, Douglas Dockery, and Arden Pope,[4] who almost invariably find that air pollution causes health problems. Yet time and again, when industry money supports an outcome industry likes, the media are careful to point this out. When industry money or EPA money supports a study whose outcome the EPA likes and industry dislikes, the source of funding goes unmentioned.

All in the Family

It is widely recognized that the EPA went well beyond what was required by the ALA lawsuit—giving rise to suspicion that it was a sweetheart lawsuit. In other words, the EPA defending itself from the ALA was comparable to Br'er Rabbit's begging his captors not to throw him into the horrible briar patch—which, it so happens, was Br'er Rabbit's home. But it is less well known that the EPA regularly funnels money to the ALA and other groups that sue it. Journalist John Merline, writing in *Investor's Business Daily (IBD)* and basing his figures on the Federal Assistance Awards Data System, noted that between 1990 and 1995, the EPA gave the ALA's national office and various state chapters more than $5 million. Yet, just between 1993 and 1996, the group filed five suits against the EPA. According to Merline, the EPA gave the NRDC over half a million dollars in 1995, even as the NRDC filed no fewer than thirty-four suits against the agency in 1993–1996. Indeed, the EPA even forked over $150,000 to the NRDC to help defray the cost of the group's legal fees for suing the agency.

"If you think the EPA is upset" about these suits, Scott Segal, a professor of environmental management at the University of Maryland, told Merline, "Think again. Truth be known, the EPA wants to be sued, because every time they are sued it expands the reach of the Clean Air Act."[5] It also expands the EPA's domain in general.

Both the ALA and the NRDC fired off letters to *IBD*, claiming that the vast majority of their EPA funding was not for purposes of suing the agency.[6] No doubt the little line on the lower left-hand corner of the EPA checks did not have the notation "to sue our pants off." But a dollar is a dollar. Sending a few million greenbacks to either group for any purpose allows them to increase funds used for other purposes, including lawsuits.

Groups like the NRDC and the EWG have been extremist since their inception. Not so the ALA, the nation's oldest charity. But sadly, it appears to have gone the way of so many civil rights and women's rights groups—once they accomplished their original goals, they felt pressed to set newer, more radical ones constantly to justify their continued existence and sustain their large budgets. Indeed, in recent years, the ALA has taken $150,000—which it uses to wage a campaign against cement kilns used to burn hazardous waste—from a group called the Association for Responsible Thermal Treatment. The group is made up entirely of commercial hazardous waste incineration companies, which is to say, the competitors of the cement kilns.[7]

The ALA also helped fill its coffers and please the EPA by campaigning for the aforementioned Envirotest-run centralized automobile emission testing program in Pennsylvania. This included full-page ads in major newspapers. Such ads are expensive, but not to worry. In the previous three years, the ALA had accepted $200,000 in contributions from Envirotest.[8] When divulged, this fact outraged many Pennsylvania ALA donors.[9] ALA lobbyist Paul Billings brushed aside criticism: "You have to make money where you can" in the charity business.[10]

In addition to handing out taxpayer money to groups that sue it, the EPA also greases the palms of groups who lobby for the agency's agenda. For example, the Ozone Transport Assessment Group, which is private but has federal sponsorship, last August announced in a memorandum that the EPA would provide members $100,000 "to support our activities" including "public service" announcements.[11]

But probably the most effective lobbying has come from the ALA and NRDC in the form of their aforementioned studies, which call for much tighter standards than even the EPA has enacted. The ALA's *Gambling with Public Health II* report, in pushing the ALA-proposed standard of 18 micrograms of particulates per cubic meter rather than the EPA's 50, essentially boasts that its standard would put so many counties out of attainment that the population in them would number 178 million, compared with "only" 85 million Americans in nonattainment counties under the EPA's proposed standard.[12] The report comes with a striking map of the United States (see figure 12–1) in which "unprotected" states are blood red, with "potentially pro-

FIGURE 12-1

AMERICAN LUNG ASSOCIATION DEPICTION OF AREA OF THE UNITED STATES UNLIKELY TO ATTAIN EPA PARTICULATE MATTER STANDARDS

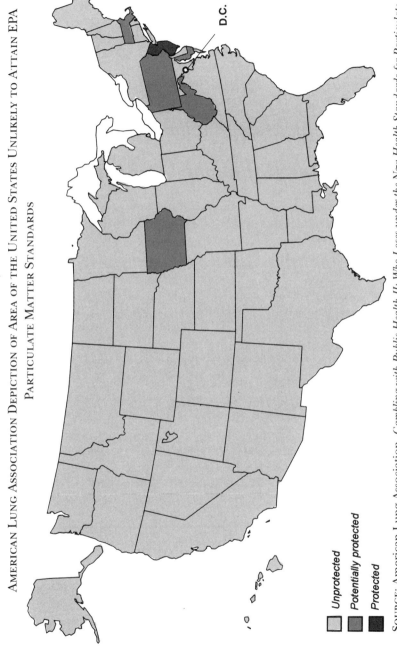

D.C.

Unprotected

Potentially protected

Protected

SOURCE: American Lung Association, *Gambling with Public Health II: Who Loses under the New Health Standards for Particulate Matter?* (Washington, D.C., January 1997).

tected" ones yellow, and "protected" green. The map is a sea of blood, with only one green state and six states plus the District of Columbia in yellow. Yet the report makes almost no effort to argue that these people really are at risk. Instead, its seventy-two pages comprise mostly a county-by-county list of the United States with the numbers of people in those counties suffering from lung or heart disorders. Out of those pages, only a few sentences so much as argue that PM2.5 is a special problem.

Then, in March 1997, the ALA officially submitted to the EPA comments on the particulate proposal. Although the comments contained ninety-four endnotes, not a single negative study was ever mentioned—none of the reevaluations of Utah, Birmingham, Philadelphia, or Steubenville; none of the European studies in the *Journal of Epidemiology and Community Health;* none.[13]

Unlike the ALA's *Gambling with Health,* the NRDC's *Breath-Taking* report has a substantial amount of text. It also boasts a lofty review board of scientists and bureaucrats. The NRDC had Joel Schwartz on hand to boost it at the May 1996 press conference announcing the report's release. But then, why shouldn't he have done so when the report states clearly it was based on Schwartz's work? Indeed, in the acknowledgments, the first two persons acknowledged—in a paragraph all their own—are Joel Schwartz and Arden Pope, two-thirds of the Particle Hunter Triumvirate.[14]

It is doubtful that either the NRDC or the ALA thinks it can get the EPA to go with the standards it proposes. It is also doubtful that this is the purpose. Rather, the real purpose is probably to push the EPA away from science and toward unsupported but emotional appeals.

"There's terrific pressure on [EPA] regulators because these groups say there will be almost 70,000 extra deaths a year if the EPA does nothing," says UC Irvine's Robert Phalen.[15] Conversely, there is every reason to think the people at the top of the EPA have no desire to withstand such pressure. After all, reports like the ALA's and the

NRDC's certainly bolster the EPA's claim that "some will contend that we are acting precipitously and others will claim that we are not being protective enough."[16] Thus, the EPA can appear in the moderate, middle position. This appearance also gave Schwartz the license he felt he needed to make the astonishing claim in a letter to the *Wall Street Journal* that "the EPA is not an interested party" in the debate over the agency's own new standards. Rather, wrote Schwartz, "it is the government agency created to judge and balance the competing claims of industry and environmental groups."[17]

The ALA- and NRDC-proposed standards also prepare the way for some date in the next century when the EPA can call once again for ratcheting down the amount of pollution allowed.

13

The Clinton Administration
Is Revolting

So industry does not like the new regulations—what else is new? And many scientists think the regulations are unsubstantiated. There is not much new about that, either.

Yet the width and depth of disgust with the EPA have even prompted heated objections and outright disgust throughout the Clinton administration. In November 1996, before issuing its promulgations, the EPA solicited comments from other branches of the administration. Many of these views did not come to light until a leak in March. The EPA had gone ahead with its standards even though practically every part of the executive branch expressed severe disagreement with at least sections of them. According to the Associated Press, these concerns were "only slightly less intense than the criticism from industry groups, members of Congress and state officials."[1] Among the objectors were persons on the President's Council of Economic Advisers; John Gibbons, the White House science adviser; staff of the Commerce, Transportation, Treasury, and Agriculture Departments; and officials at the Small Business Administration. Indeed, only the Interior Department gave complete approval to the EPA proposal, basing its position on evidence that "plants are more sensitive than humans to ozone."[2]

Assistant Secretary of Transportation Frank Kruesi, for example, wrote that it was "incomprehensible that the Administration would commit to a new set of standards and new efforts to meet such standards without much greater understanding of the problem and its solutions."[3] Other specific objections, such as those from the Small Business Administration and the Council of Economic Advisers, have been noted.

Some objectors said that not enough scientific work had been done. "I find it hard to believe that we would suffer more than we would gain by taking more time for further interagency review, consensus building and additional analysis," Gibbons wrote.[4] Others said the science simply did not support the EPA position. "Current data do not support clear associations of [premature mortality] effects with either fine particles (PM2.5), inhalable particles PM10 or PM15, [or] sulfate, so that causality for the observed mortality and morbidity effects cannot be established," Rosina Bierbaum, acting associate director of OSTP, wrote in a November 1996 memorandum. She added that "the database for actual levels of PM2.5 is also very poor, and only a handful of studies have actually studied PM2.5 effects, per se."[5]

The Katzen Coverup

The rebellion did not end with the November memorandums, nor did the EPA's efforts to squelch it. In December 1996, House Commerce Committee Chairman Thomas Bliley asked the Office of Information and Regulatory Affairs (OIRA), a section of the Office of Management and Budget (OMB), for a reading on the EPA's handling of the air pollution proposal. OIRA is supposed to review economic analyses that agencies are required to prepare indicating that they have regulated in the most cost-effective manner possible. To help this procedure along, OIRA has prepared a best-practices document to identify procedures that will provide the best analysis of the impact of a regulation. On January 15, 1997, the committee received a document that was generally supportive of EPA. But this was a revised draft. What had appeared in the two earlier drafts, both drawn up in early January, had horrified EPA officials.

The key sentence of the original twenty-seven-page (double-spaced) document read, "While these analyses produced much useful information, there were several areas in which they did *not fully conform* to the principles discussed in the Best Practices document" (emphasis added). OMB's general counsel suggested that the response be shortened, and it was, to about fifteen pages. But it retained, word for word, the damning sentence. Thus, after reviewing the second draft, an EPA deputy director within the Office of Air and Radiation, John Beale, wrote to the OIRA economist in charge, Art Fraas, complaining the response placed too much emphasis on differences between the agencies and "could be very damaging" to the rule change.[6] But Beale did not stop there. He repeatedly telephoned Mary Nichols, his direct boss, to urge her to apply pressure to OIRA director Sally Katzen. Nichols did just that. Under pressure from Beale, Katzen, and Nichols, Fraas finally gave in.[7]

The final draft of the report was snipped down to a mere twelve pages. More important, the key sentence had been changed so that it now read, "While these analyses *were consistent with* the Best Practices document and produced much useful information, there were several areas where additional work would have been productive" (emphasis added).[8]

Bliley was incensed when he found out about the intrigue that surrounded the OMB report. Such action amounted to "exerting extraordinary control" over an office "that is supposed to review federal agencies' regulatory actions," Bliley complained in a letter to Browner.[9]

Yet, that was not the only favor OMB was to do for the EPA. The November memorandums were not formal responses to the EPA; the formal ones were to be submitted in March 1997 under the Administrative Procedure Act. For the most part, they were not. To be sure, much work went into them. Fred Lipfert prepared a fifty-page report for the Department of Energy.[10] Yet, the deadline came and went, and the comments were not turned in. "The comment period for the proposals closed Wednesday, March 12, 1997. No executive branch departments or agencies, including USDA, submitted written comments to the docket," an undated Agriculture Department memo later stated.[11] In fact, a few agencies had made submissions, including—strangely enough, considering the origin of the memo—the USDA. Among the official comments filed was the Pentagon's, which said that "the establishment of a PM2.5 [standard] as proposed would critically impact its ability to properly train military personnel," the "Army and Marine Corps are already suffering training constraints," and services might have to "limit military aircraft operations" even under the *current* regulations.[12]

But most comments were not submitted. Why not? OMB stopped them. A March 11, 1997, e-mail from Jean Vernet of the Department of Energy's Office of Global Environment to other staff members within the department stated: "Based on reports from a meeting this morning with [OIRA administrator] Sally Katzen, at which Dan Reicher and Kyle Simpson represented the Department, Federal agencies will not [*repeat not*] be transmitting comments on the EPA O3/PM [ozone and fine particle] proposals. To what extent agency comments will be entertained during a yet undefined interagency review effort is unclear" (emphasis in original).[13]

"All the [department] secretary-level comments came into OMB, and there was so much discord and problems with the new standards [they proposed] that they did not want to show the administration was so fractured," one USDA official told me on the condition of anonymity. According to him, Katzen "said we're not going to show these fractured efforts within the administration." The anonymous official added that, "instead of handing in written documents, they've established some kind of interagency working committees to iron out problems. But they will be behind closed doors."

His view is supported from notes that I obtained from a November 11, 1996, interagency briefing on the proposed standards, written on letterhead of the OSTP, which is part of the executive branch. "NO meetings with outsiders on this matter!!" it states. "If someone asks for a meeting, refer them to Sally Katzen. No one who talks to their constituents about these matters will be invited back to future meetings—this MUST be run right [as] it is a 'poster child' for reg reform" (emphasis in original).[14]

Katzen declined to let me interview her on this subject.[15]

14

First Do No Harm

Fundamental to the position of the EPA and others advocating PM2.5 regulation is that, regardless of whether PM2.5 per se may or may not be a direct cause of health problems, the association appears strong enough that directly controlling PM2.5 is still warranted. That way, what is causing harm can be reduced, even if the cause is not exactly known.

Curiously, Carol Browner seems ignorant of this. During congressional testimony in May 1997, Rep. Thomas Sawyer (D-Ohio) put her on the spot by asking if there were chemical differences among PM2.5 particles. "I'm not sure I understand the question of chemical differences," she replied. Mary Nichols, sitting at her left, had to answer for her that "all PM2.5 is not the same." But Browner continued to step on her tongue. "Are you telling me there's no difference regardless of what kind of stuff it is?" Sawyer asked. "Right," said Browner. "It is the fine particle that causes the problem . . . . It is the size."[1]

Au contraire, says the EPA's own staff paper. It states that "identification of specific components and/or physical properties of fine particles which are associated with the reported effects is very important . . . for reducing health risks."[2] Recall that the ALA's report stated, "It is not known whether the size and concentration of the inhaled particles are wholly responsible for the health effects, or if their chemical composition also plays a role." This is the position of the very scientists whose work Browner cites. In one of the studies she discussed at a congressional hearing, the authors concluded, "These results suggest that what a particle is made of is more important to human health than the particle's mass, per se."[3] Arden Pope certainly would not give Browner his blessing. "I'm not pretending I know precisely what it is in these particles that causes the damage," he told me. "I can't tell you whether it's because of their chemical makeup or because their size makes them more of an irritant."[4]

Aside from the confused EPA administrator, the advocates of PM2.5 controls are saying that even if the size of the particle per se does not count, reducing PM2.5 will still reduce whatever nasties out there are causing the problem. By ignoring for the moment that so many of the PM2.5 studies cited do not even show an association with illness, much less causation, the assertion would not be wholly illogical. Throughout history, people have put together associations between diseases and possible causes and benefited from their guesses, even when the guessed-at cause was sometimes wrong. The ancient Romans, for example, blamed malaria on the bad air (*mal* and *aria*) coming up from swamps and therefore avoided low-lying areas in the summer. By so doing, they avoided the real culprit—the disease-carrying mosquitoes infesting the swamps.

Conversely, guessing at associations has also had devastating consequences. During the Great Plague of London in 1665, the citizenry got the idea that the disease was being spread by animals—good thinking. So they killed all their cats—bad thinking. The disease was being spread by the lice of rats, which the now-dead cats could no longer kill.[5]

Thus, there are three possible outcomes if PM2.5 is used as a surrogate for something else in the air that is bad but unknown. First, it is a proper surrogate. Reduce the PM2.5 and the harmful stuff is also reduced. Second, there is no effect at all on reducing the bad stuff.

"I'm an engineer," says K. C. Shaw, at the Geneva Steel Mill in Utah County. "I put air pollution control devices on things in my plant. But if you tell me to reduce PM2.5 concentrations, I want to know what is the causal agent. Is it the size, is it the number of particles that are produced, is it what they're made out of—the chemical constituency—is it the surface area, is it what's absorbed onto the particle?"

Shaw continued: "I could install a tremendous number of devices and have absolutely no impact on health. So to simply tell me to reduce PM2.5 is far too broad. I know that sounds like obstructionist or delay tactics, but it's not. I have to know more about the causal agent to know what to do."[6]

Then there's what's behind door number three. The third possibility is that we could actually be making things worse.

When it comes to environmental promulgations, the rule often blazes the path for the science, rather than the other way around. If there is anything in air pollution that is causing harm and it is not PM2.5, the new regulations could delay discovering what it is for years. We have already seen this happen with PM10: the EPA's regulation led to scores of PM10 studies both in the United States and abroad but only a handful that measured PM2.5. "What we're doing now is simply going to set up a second regulatory lamppost," says Roger McClellan, alluding to the joke about the drunk who said he was searching for his keys under the lamppost not because he thought he had dropped them there but because that was where the lighting was better. "We'll have pre-ordained what we have data on,"[7] Frederick Lipfert agrees. "The EPA will have mandated monitoring of PM2.5 and that means you'll be able to evaluate the PM2.5 hypothesis but not any other," he says.[8]

Another possibility is that, by enforcing PM2.5 rules, we could actually be pushing polluters toward creating particles that are more dangerous. "For example," says Robert Phalen, "if it's the ultrafine particles that are highly toxic," by which he means those considerably smaller than PM1.0, "the [PM2.5] regulation could lead to more deaths than lives saved." Diesel engines, he notes, historically have emitted particles in the PM2.5 range. In late 1980s, at the urging of the EPA, new engines were designed that put out particles smaller in size. "They put out less mass, but they put out thirty to sixty times more particles by number. If the studies indicating that ultrafine particles are the real hazard are correct, changes such as these would make the air more hazardous even as they brought you closer to compliance with the [EPA] standards," says Phalen.[9]

Such an outcome is far from certain. But the mere possibilities show how the law of unintended consequences can be at work in the sanctified environmental efforts of the EPA as well as in other places. "If mass is not the right measure of toxicity," Phalen explains, "then control strategies are misdirected, and the results could be catastrophic."[10]

Folly at a Glance

Here in a nutshell is what the science really appears to show about particulates:

1. Particulates are often associated with increased deaths and hospitalizations, but often they are not. The same goes for numerous other pollutants.

2. To the extent that there is an association with both deaths and hospitalizations, it is a consistently weak one— deep in the range that epidemiologists generally treat as having no meaning.

3. To the extent that there is an association, it is just that, an association. Ambient particle levels may be a surrogate for adverse weather or some other health threat such as indoor allergens. There is a pittance of toxicological or clinical data supporting particles themselves as a cause of

sickness or death. The cause may be a specific chemical or something else that will be overlooked in the stampede to regulate PM2.5.

4. Even if particulates were causing premature deaths and hospitalizations, it is far from clear that PM2.5 is more closely associated with sickness and death than PM10, which is already stringently regulated.

Here in a smaller nutshell is what the science appears to show about ozone:

1. It is not apparently killing anybody at the levels found in the United States and may actually be saving lives by preventing skin cancer.

2. It does cause a decrease in lung function, to the extent that extremely sensitive individuals in extreme conditions may suffer enough discomfort to go to the hospital.

3. Even if there were no man-made ozone at all, naturally present ozone would probably still be enough to cause these problems in a certain portion of the population.

Given all this, the case for burdening Americans with perhaps major lifestyle changes and scores of billions of dollars each year for new regulations seems rather absurd. It is all the more so considering that the levels of these pollutants are already much lower than they have been and are still dropping.

EPA figures released in December 1996 showed that all six of its target pollutants, including ozone and particles, have been declining steadily for ten years. The measured pollutants have decreased nationally by almost 30 percent, even as the population has increased by 28 percent and vehicle miles increased by 116 percent. Ozone levels have fallen by 6 percent and would have fallen more except for, as even the EPA notes, unusual weather conditions in 1995.[11]

As for PM10, just between 1988 and 1995, these concentrations nationwide fell 22 percent, while emissions dropped 17 percent. This trend is not slowing; indeed, just from 1994 to 1995, PM10 emission levels dropped another 6 percent while concentration levels fell by 4 percent. Although PM10 was not measured before 1988, there is every reason to believe that it was already far lower than it had been. Back when the EPA was measuring suspended particulates as a whole (of which PM10 is a large proportion), these dropped more than 20 percent over the decade. Average particle levels in U.S. cities are now perhaps one-fiftieth of those in London in the 1950s, when it suffered its great killer fogs. Likewise, as the EPA's own particulate staff paper points out, PM2.5 concentrations will continue to decrease because of the implementation of existing control programs that target PM10 and PM2.5 precursors.[12]

In other words, there are far fewer particulates of all sizes in the air than there were ten years ago, far fewer yet than ten years before that, and fewer still than the levels known to cause death in London. Further, without a single new rule, a single new sentence uttered from the mouth of Carol Browner, a single new alarming news story, a single heart-rending TV ad from the Clean Air Network, or a single new scary report from the ALA or NRDC, less PM10 and hence less of the PM2.5 subset will be emitted. Less PM10 and PM2.5 will be in the air. Less will be breathed. All this is true for the gases that form ozone as well. The best that can be hoped for from the new standards is that trends already in place might be accelerated. Yet even the EPA appears to concede there might not even be much acceleration, considering that its cost-benefit analyses for both ozone and particulates assume that many regions of the country cannot comply with their proposed new laws.

In the meantime, the lack of new regulation gives scientists more leeway in being creative about what pollutants they measure and study. Further, it is not as if there will not be money to do so—the fiscal 1998 budget calls for $26.4 million for such studies.[13] "Let's take a big time out," CIIT's Roger McClellan suggests. "Why not reaffirm present PM standards and see what happens?" For now, he said, "The scientific basis for PM2.5 is just too weak and we really run a risk of prematurely identifying a new target absent of understanding what that target should be."[14]

Already, the EPA has switched from saying that total particulates are the real problem to saying PM10 is the real problem to saying PM2.5 is the real problem. What if the EPA is wrong yet again?

A reasonable solution, it seems, is to step up the PM2.5 monitoring system, including more measurements of the chemical composition of the PM2.5, not just the size alone. Since the enforcement part of the EPA's new standards does not begin until the year 2002 anyway, this would not necessarily put us that much behind schedule. Scientists such as UC Irvine's Daniel Menzel say that we also need to begin looking more at pollution interactions, rather than at single pollutants.[15]

15

An End Run around Science and Reality

In February 1997, Carol Browner told a Senate panel, "This has been the most extensive scientific process ever conducted by EPA for public health standards."[1] As we have seen, the scientific process, properly defined, played almost *no part* in the EPA's decision except as something to be deftly worked around. We have also seen that Browner is largely ignorant of the science. She has apparently not read any particulate studies. She does not know whether the Clean Air Scientific Advisory Committee reviewed eighty-six or eighty-seven studies, except that whichever number it is, it is important enough to bear repeating ad nauseum. She does not know that even the scientists who call for PM2.5 regulation do not claim that the size of the particle per se causes health problems. She has not even gone to the bookstore to get *Air Pollution for Dummies*.

So what *did* prompt this move that Joseph Lyon called closer to witchcraft than to science? It might be explained by public choice theory. Carol Browner may not know much about particulates or ozone, but she knows that she is the head of a regulatory agency and regulators regulate. And enforcing old regulations is not nearly as fulfilling as proposing new ones. More regulations and lowering the boom on emissions, to her, are Very Good Things. In this sense, Browner's and the EPA's actions not only are not unusual but are predictable. (Also predictably, if they get what they want this time, at some point they will find a justification to get even more.) What makes this case particularly important is that it may be the single best example of a powerful federal agency imposing a crushing regulatory burden sheerly for the sake of regulating.

Conversely, Robert Hahn, resident scholar at the American Enterprise Institute, thinks that the proposed standards are a bureaucratic power grab run amok. Any area that is out of attainment is to a great extent at the agency's mercy for building permits and uses, transportation plans, industrial uses, and the like. "The EPA is likely to use this power not to shut the country down but to allow the bureaucrats and political appointees to curry favor according to their interests. If you have a project they like, they'll reward you. And if not, they won't reward you," Hahn observed.[2]

That may sound a bit paranoid, but it's not entirely unreasonable, considering that for the first time the EPA has promulgated standards that its own cost-benefit analyses (by looking only at partial attainment) seem to be conceding cannot be met. It is also hinted at in a memo that Assistant Secretary of Transportation Frank Kruesi wrote in November 1996, when he complained that the proposals would "bring a significantly larger proportion of the population and more jurisdictions under Federal oversight and procedural burdens."[3] Perhaps more than one person at the EPA read this memo, shook his head, and thought, "That's just the point, you idiot!"

Too, the arrogance that Browner and the EPA have shown on this issue clearly reflects a predictable attitude of an agency given almost complete free reign by both Congress and the president for a quarter of a century. Until recently, what the EPA promulgated became law—no fuss, no muss, no messy papers for presidents to sign. That has changed in recent years because of three pieces of legislation. One, the Congressional Review Act of 1996, effectively gives Congress the power to "veto" agency regulations. It has sixty days after the EPA's final promulgation to do so.[4] Congress may also exercise such a "veto" under the Small Business Regulatory Enforcement Fairness Act (SBREFA), discussed earlier.

SBREFA also allows private parties to sue the EPA if they feel the act has not been complied with. Indeed, almost simultaneously with the EPA's promulgation of the final regulations, the National Trucking Association, the U.S. Chamber of Commerce, and several smaller groups filed suit in federal court to block them.[5]

The final provision, the Unfunded Mandates Reform Act passed in 1995, also allows lawsuits. It states that any regulation not dealing with national security and costing the private sector more than $100 million in any year must have congressional approval.[6] The EPA standards may be tied up in the courts until the next century.

Whether Congress will invoke its "veto" or will allow the EPA to hurtle on like a semitrailer with broken airbrakes cannot be determined. Certainly, the Gingrich Revolution was not too revolutionary on environmental issues, notwithstanding that during the last campaign the Democrats fired blasts at what they called the *E. coli* Republicans—meaning they were soft on health issues.[7]

Two letters sent to President Clinton do offer hope, though. The first, by Rep. John Dingell (D-Mich.) with forty-one other Democratic signatories, does not oppose the new standards per se but expresses concern that they "will create such controversial and impractical targets that they will undermine support for the [Clean Air] Act, even amongst its friends"[8] and calls for a meeting with the president and Vice President Al Gore on the subject. The second goes even further. Written by Rep. Rick Boucher (D-Va.) and signed by thirty-five other Democrats and seventy-eight Republicans, it declares, "There are serious scientific uncertainties regarding the health benefits associated with the [new] regulations" and "the significant uncertainty surrounding the health benefits, costs and effects of the [new] ozone and fine particulate standards requires that the imposition of the regulations be deferred."[9]

As to the motivation of the supporting players, the environmentalist groups, it is simple to conclude that they have gone from looking for pitchforks in haystacks to searching for needles because nobody wants to put themselves out of work—or watch contributions fall off. These are crisis-based organizations.

But a final factor seems to reflect the second great shift in environmentalist strategy in the past few years. The first shift was a move away from assertions that could ultimately be proved (and thus ultimately disproved) toward those that are nonfalsifiable. "This is the case" has increasingly been replaced with "what if it were the case?" This shift was necessitated by a realization that science would never support the more extreme claims.

The second shift was caused by a realization that society will not buy into environmentalist dogma on what *pollution* constitutes. The environmentalist effort has failed to convince us that all man-made emissions everywhere are pernicious, regardless of whether they cause harm to health or even aesthetics. Americans do not buy it—literally. The number of Americans willing to pay for more environmental regulation fell from 71 percent to 40 percent in just the years from 1990 to 1994.[10] And despite Browner's soothing claims, people do worry about their lawn mowers and barbecues. Most Americans are now convinced that emissions that do not harm do not constitute pollution. The only conceivable strategy to counter this is to make us believe that *all* emissions do cause harm—and not just to trees or buildings or aesthetics but to human health. In this way, the EPA and environmentalists get the desired result anyway—anything that man puts out is a pollutant. Rules that would effectively bring pollution levels down to those of the natural background in some geographic areas—as would both the EPA's proposed particulate and ozone standards—would be a final affirmation of this.

Considering that the true practitioners of witchcraft have been worshippers of nature, the metaphor fits. And considering that the EPA's newest onerous air pollution standards are supposed to be a "poster child for regulatory reform," the future of applying science (and reason) to health and environmental laws seems dim indeed.

Mortality Studies with Direct Measurement of Fine Particles

(chronological order)

Dockery, Douglas, Joel Schwartz, and John D. Spengler. "Air Pollution and Daily Mortality: Associations with Particulates and Acid Aerosols." *Environmental Research* 59 (December 1992): 362–73. No statistically significant association was found between deaths and fine particles in either of the two areas analyzed, St. Louis or Eastern Tennessee.

Dockery, Douglas W., et al. "An Association between Air Pollution and Mortality in Six U.S. Cities." *New England Journal of Medicine* 329 (December 9, 1993): 1753–59. This was the Six Cities study. It found a statistically significant 26 percent increased risk of death for all persons when comparing the least polluted city with the most polluted one. But among persons claiming never to have smoked, it found only a nonsignificant increased risk. It also found no significantly increased risk among those with "no occupational exposure to gases, fumes, or dust."

Pope, C. Arden III, et al. "Particulate Air Pollution as a Predictor of Mortality in a Prospective Study of U.S. Adults." *American Journal of Respiratory and Critical Care Medicine* 151 (March 1995): 669–74. Commonly called ACS II, this study found a statistically significant association between deaths in the least polluted of fifty cities and the most polluted city, with 17 percent more deaths among all persons and 22 percent more deaths among persons who claimed never to have smoked.

Schwartz, Joel, Douglas W. Dockery, and Lucas M. Neas. "Is Daily Mortality Associated Specifically with Fine Particles?" *Journal of the Air & Waste Management Association* 46 (October 1996): 927–39, especially table 5. This was the follow-up on the Six Cities study. It looked at each city to detect an association between death and an increase of 10 micrograms per cubic meter or more in fine particles. In three cities, it found a statistically significant association; in three, it did not. One of the negative cities, Steubenville, also had the largest increase in fine particles. But the large size of the dataset from one of the significant cities (Boston) made the association in all the cities combined statistically significant.

Illness Studies with Direct Measurement of Fine Particles

(chronological order)

Dockery, Douglas W., et al. "Effects of Inhalable Particles on Respiratory Health of Children." *American Review of Respiratory Disease* 139 (March 1989): 587–94. This was a child study, using data from the Harvard Six Cities study. It found no statistically significant increase of respiratory symptoms for all three classes of children studied: those with bronchitis, chronic cough, or chest illness. It was further broken down by those with or without wheezing or asthma. Again no statistically significant increases were found.

Ostro, Bart D., et al. "Asthmatic Responses to 'Airborne Acid Aerosols." *American Journal of Public Health* 81 (June 1991): 694–702. This was a child study. It found no statistically significant association with the three symptoms looked at: cough, shortness of breath, or asthma ratings.

Thurston, George D., et al. "Respiratory Hospital Admissions and Summertime Haze Air Pollution in Toronto, Ontario: Consideration of the Role of Acid Aerosols."

Fine particle measurement was PM2.1 for Neas et al. (1995), Neas et al. (1996), Dockery et al. (1996), Raizenne et al. (1996), and Delfino et al. (1997).

Environmental Research 65 (May 1994): 271–90. This study found a statistically significant association between hospital admissions and three types of pollutants, but not fine particles.

Schwartz, Joel, et al. "Acute Effects of Summer Air Pollution on Respiratory Symptom Reporting in Children." *American Journal of Respiratory and Critical Care Medicine* 150 (November 1994) 1234–42. This was a child study. It found a statistically significant increase in coughing associated with fine particles in general. Otherwise, the study concluded, "There was no evidence that other measures of particulate pollution. . . were preferable to PM10 in predicting incidences of respiratory symptoms." Thus, it proffered no evidence for supporting a PM2.5 standard.

Neas, Lucas M., et al. "The Association of Ambient Air Pollution with Twice Daily Peak Expiratory Flow Rate Measurements in Children." *American Journal of Epidemiology* 141 (January 15, 1995): 111–22. This was a child study. It found a statistically significant association between fine particles and coughing, but none with wheezing or cold symptoms.

Neas, Lucas M., "Fungus Spores, Air Pollutants, and Other Determinants of Peak Expiratory Flow Rate in Children." *American Journal of Epidemiology* 143 (November 8, 1996): 797–807. This was a child study. It found a statistically significant association with fine particles, especially strongly acidic ones, for coughing. For reduced breathing flow, it also found such an association with the strongly acidic particles, but not for fine particles in general.

Dockery, Douglas W., et al. "Health Effects of Acid Aerosols on North American Children: Respiratory Symptoms." *Environmental Health Perspectives* 104 (May 1996): 500–505. This was a child study. It found a statistically significant association between bronchitis and fine particles from sulfates. No associations were found with any of the other symptoms considered, including asthma, wheezing, coughing, or phlegm production. No association was found with nonsulfate fine particles.

Raizenne, Mark, et al. "Health Effects of Acid Aerosols in North American Children: Pulmonary Function." *Environmental Health Perspectives* 104 (May 1996): 506–14. This was a child study. It found a statistically significant association with fine particles, especially strongly acidic ones, for a variety of respiratory functions.

Delfino, Ralph J., et al. "Effects of Air Pollution on Emergency Room Visists for Respiratory Illnesses in Montreal, Quebec." *American Journal of Respiratory Critical Care Medicine* 155 (February 1997): 568–76. This study found no statistically significant association with fine particles and emergency room admissions.

The Studies Listed on Browner's Poster of February 12, 1997

(chronological order)

Schwartz, Joel, Douglas W. Dockery, and Lucas M. Neas. "Is Daily Mortality Associated Specifically with Fine Particles?" *Journal of the Air and Waste Management Association* (October 1996): 927–39, especially table 5. The poster claims the adverse health effect of PM2.5 is "premature mortality." The results for the three cities with statistically significant associations between PM2.5 and premature death were listed—fair enough. But how about those three cities that showed no such association? They were left off the chart.

Thurston, George D., et al. "Respiratory Hospital Admissions and Summertime Haze Air Pollution in Toronto, Ontario: Consideration of the Role of Acid Aerosols." *Environmental Research* 65 (May 1994): 66–74. The poster claims the adverse effect of PM2.5 is "hospital admissions." Thurston and colleagues did not find a statistically significant association with fine particles. Moreover, they stated, "these results suggest that what a particle is made of is more important to human health than the particle's mass, per se."

Schwartz, Joel, et al. "Acute Effects of Summer Air Pollu-

tion on Respiratory Symptom Reporting in Children."
American Journal of Respiratory and Critical Care Medicine 150 (November 1994): 1234–42. The poster claims
the adverse effect of PM2.5 is "respiratory symptoms."
As noted in appendix A, that study actually concluded
that "there was *no* evidence that other measures of particulate pollution" were preferable to PM10 (emphasis added). Thus it proffers no evidence for supporting
a PM2.5 standard.

Pope, C. Arden III, et al. "Particulate Air Pollution as a Predictor of Mortality in a Prospective Study of U.S.
Adults." *American Journal of Respiratory and Critical Care
Medicine* 151 (March 1995): 669–74. This was the ACS
II study. The EPA poster accurately presented its conclusion tying particulates to premature death. Nevertheless, the study looked only at fine particles, so it could
not tell if other pollutants or larger particles might have
been responsible.

Dockery, Douglas W., et al. "An Association between Air
Pollution and Mortality in Six U.S. Cities." *New England
Journal of Medicine* 329 (December 9, 1993): 1753–59.
The poster claims the adverse effect of PM2.5 is "premature mortality." Premature deaths were found only
in persons who had been smokers or had been exposed to occupational air pollution.

Notes

PREFACE

1. Stanley Davis, mayor of Salmon, Idaho, personal telephone communication with the author, May 8, 1997.

2. William H. Miller, "Clean Air Contention," *Industry Week*, May 5, 1997, p. 14.

3. 61 FR 65638, Proposed Rules, Environmental Protection Agency (EPA), 40 CFR Part 50, [AD-FRL-5659-5] RIN 2060-AE66, National Ambient Air Quality Standards for Particulate Matter: Proposed Decision, Part II, Friday, December 13, 1996, Action: Proposed Rule, *Federal Register*, vol. 61; 61 FR 65716, Proposed Rules, Environmental Protection Agency (EPA), 40 CFR Part 50, [AD-FRL-5659-4] RIN 2060-AE57, National Ambient Air Quality Standards for Ozone: Proposed Decision, Part III, Friday, December 13, 1996, Action: Proposed Rule, *Federal Register*, vol. 61, no. 241.

4. Press Release, EPA, "EPA Proposes Air Standards for Particulate Matter and Ozone," November 27, 1996.

5. "National Ambient Air Quality Standards for Particulate Matter," 62 FR 38762, July 18, 1997; 62 FR 38856, July 18, 1997. See also, "Presidential Comments" at 62 FR 38421, July 16, 1997.

CHAPTER 1: SOMETHING IN THE AIR

1. Clean Air Act of 1970 and amendments, 42 U.S.C.

2. See EPA, *Review of the National Ambient Air Quality Standards for Particulate Matter: Policy Assessment of Scientific and Technical Information OAQPS Staff Paper* (EPA-452/R-96-007, July 1996), pp. II-3 and IV-IV.

3. American Lung Association v. Browner, CIV-93-643-TUC-ACM (D-Ariz., October 6, 1994).

4. Jack Gibbons, Office of Science and Technology Policy, memorandum to Sally Katzen, November 21, 1996.

5. For example, Stephen B. Huebner and Kenneth W. Chilton, *More than a Particle of a Doubt: The Science behind EPA's*

Particulate Proposal, Policy Brief 178 (St. Louis: Center for the Study of American Business at Washington University, April 1997), p. 2.

6. See EPA, *Review of the National Ambient Air Quality Standards for Particulate Matter,* pp. II-1–2.

7. See Michael Fumento, "The Golden State: Trendsetter or Tocsin?" *ECO,* vol. 1, January 1994, p. 32.

8. Frank Reeves, "EPA Hardens Fumes Stance," *Pittsburgh Post-Gazette,* October 7, 1994, p. B1.

9. Frank Reeves, "Emission Lawsuit Settled," *Pittsburgh Post-Gazette,* December 7, 1995, p. A1.

10. *Health and Environmental Effects of Particulate Matter,* U.S. EPA Fact Sheet, November 29, 1996.

11. Its original press release actually mentioned a 40,000 figure, but close reading reveals that this was the agency's estimate for *all* pollution-related deaths, not those that would be prevented with the proposed new regulations.

12. *Health and Environmental Effects of Particulate Matter.*

13. Deborah Sheiman Shprentz et al., *Breath-Taking: Premature Mortality Due to Particulate Air Pollution in 239 American Cities* (New York: Natural Resources Defense Council, May 1996), p. 1.

14. See Michael Fumento, *Science under Siege* (New York: Morrow, 1993), pp. 27–44.

15. American Lung Association, "Comments of the American Lung Association to the U.S. Environmental Protection Agency Regarding Proposed Decision: National Ambient Air Quality Standards for Particulate Matter (61FR65638)," March 12, 1997, p. 2; Shprentz, *Breath-Taking,* pp. 129–30.

16. American Lung Association, *Gambling with Public Health II: Who Loses under New Health Standards for Particulate Matter* (Washington, D.C.: American Lung Association, January 1997), p. 7, table 1.

17. ALA, "Comments of the American Lung Association to the U.S. Environmental Protection Agency," p. 2.

18. Shprentz et al., *Breath-Taking,* p. 5.

19. See, for example, her column, Frederic Perera, "Green-Bashing: A Health Hazard," *New York Times,* July 29, 1993, p. A23; see "Group Calls Particulate Regulations Inadequate," Associated Press, May 29, 1980.

20. 52 FR 24854, July 1, 1987.

21. EPA, *Review of the National Ambient Air Quality Standards for Particulate Matter,* p. V-1.

22. As quoted in transcript of White House briefing, June 25, 1997.

23. See http://Tnwww.rtpnc.epa.gov/naaqspro/pmlist.htm.

24. See *Controlling Particulate Matter under the Clean Air Act: A Menu of Options* (State and Territorial Air Pollution Program Administrators/Association of Local Air Pollution Control Officials: Washington, D.C., July 1996), pp. 48–49, table 1.

25. For example, Tony Snow, "Renaissance Chance for Gore?" *Washington Times*, June 2, 1997, p. A14.

26. Pie charts showing this can be found at http://134.67.104.12 /G-DRIVE/O3PMRH/NRDIRS/PIECHRT.PDF. It only represents three cities, but that's all the EPA currently has data for.

27. Will Humble, epidemiology program supervisor with the Association of State and Territorial Health Officers, as quoted in Mary Jo Pitzl, "On Clean Air Day, Symington Assails Clean-Air Plan," *Arizona Republic*, February 28, 1997, p. A1.

28. EPA Administrator Carol Browner and EPA Assistant Administrator for Air and Radiation Mary Nichols, press conference, "Regarding a Proposal to Strengthen National Air Quality Standards," EPA Headquarters, November 27, 1996.

29. Carol Browner, EPA administrator, prepared testimony before the U.S. Senate Committee on Environment and Public Works, February 12, 1997.

30. One example is during Browner's testimony before the U.S. House of Representatives Subcommittee on Health and Environment and Subcommittee on Oversight and Investigations, Committee on Commerce, May 15, 1997.

CHAPTER 2: HOW GRIM WAS MY VALLEY?

1. David Murray, personal telephone communication with the author, March 3, 1997.

2. C. Arden Pope, Joel Schwartz, and Michael R. Ransom, "Daily Mortality and PM10 Pollution in Utah Valley," *Archives of Environmental Health*, vol. 47 (May-June 1992), pp. 211–17.

3. U.S. Bureau of the Census, *Statistical Abstract of the United States: 1995* (Washington, D.C.: Bureau of the Census, 1995), p. 144, table 219.

4. C. Arden Pope III, "Respiratory Disease Associated with Community Air Pollution and a Steel Mill, Utah Valley," *American Journal of Public Health*, vol. 7 (May 1989), pp. 623–28.

5. Joseph Lyon, M.D., personal telephone communication

with the author, January 27, 1997.

6. Joseph L. Lyon et al., "An Every Other Year Cyclic Epidemic of Infants Hospitalized with Respiratory Syncytial Virus," letter, *Pediatrics*, vol. 27 (January 1996), pp. 152–53.

7. Pope, Schwartz, and Ransom, "Daily Mortality," p. 214, table 2.

8. Joseph L. Lyon, Motomi Mori, and Renlu Gao, "Is There a Causal Association between Excess Mortality and Exposure to PM10 Air Pollution? Additional Analyses by Location, Year, Season, and Cause of Death," *Inhalation Toxicology*, vol. 7 (July 1995), pp. 603–14.

9. Ibid.

10. Ibid.

11. Patricia Styer et al., "Effect of Outdoor Airborne Particulate Matter on Daily Death Counts," *Environmental Health Perspectives*, vol. 103 (May 1995), pp. 490–97.

12. Beverly Freeman, Office of the Dean, Harvard School of Public Health, personal written communication to the author, May 19, 1997.

CHAPTER 3: LITTLE THINGS MEAN A LOT

1. Joel Schwartz, "Air Pollution and Daily Mortality in Birmingham, Alabama," *American Journal of Epidemiology*, vol. 137 (May 15, 1993), pp. 1136–47.

2. Joel Schwartz, "Air Pollution and Hospital Admissions for the Elderly in Birmingham, Alabama," *American Journal of Epidemiology*, vol. 139 (March 15, 1994), pp. 589–98.

3. Jonathan M. Samet, Scott L. Zeger, and Kiros Berhane, "The Association of Mortality and Particulate Air Pollution," *Particulate Air Pollution and Daily Mortality* (Health Effects Institute: Cambridge, Mass., July 1995), pp. 1–104.

4. For a fascinating article on this phenomenon, see Richard Leviton, "How the Weather Affects Your Health," *East West*, vol. 19 (September 1989), p. 64.

5. Jerry M. Davis et al., *Airborne Particulate Matter and Daily Mortality in Birmingham, Alabama*, Technical Report no. 55 (Research Triangle Park, N.C.: National Institute of Statistical Sciences, November 20, 1996), p. 1.

6. Carl M. Shy, member of Clean Air Scientific Advisory Committee, prepared testimony before the U.S. House Subcommittee on Health and the Environment and the Subcommittee on Oversight and Investigations, Commerce Committee, May 8, 1997.

7. Lynn Rosenberg, "Induced Abortion and Breast Cancer: More Scientific Data Are Needed," *Journal of the National Cancer Institute,* vol. 86 (November 2, 1994), pp. 1569–70.

8. As quoted in Curt Suplee, "Higher Risk of Breast Cancer Found in Young Women Who Had Abortions," *Washington Post,* October 27, 1994, p. A3.

9. National Cancer Institute Press Release of January 1, 1995, as quoted in *Cancer Biotechnology Weekly,* January 16, 1995, p. 4.

10. U.S. EPA, "Respiratory Health Effects of Passive Smoking: Lung Cancer and Other Disorders" (EPA/600/6-90/006F).

11. Neil Roth, of Roth Associates, Rockville, Maryland, personal telephone communication with the author, March 3, 1997.

12. Joel Schwartz, "Particulate Air Pollution and Daily Mortality in Detroit," *Environmental Research,* vol. 56 (December 1991), pp. 204–13.

13. Ron Wyzga, senior manager of Air Quality Risk and Health Assessment at Electric Power Research Institute, personal telephone communication with the author, May 29, 1997. The paper is being prepared for publication.

14. Joel Schwartz and Douglas W. Dockery, "Particulate Air Pollution and Daily Mortality in Steubenville, Ohio," *American Journal of Epidemiology,* vol. 135 (January 1, 1992), pp. 12–19.

15. Suresh H. Moolgavkar et al., "Particulate Air Pollution, Sulfur Dioxide, and Daily Mortality: A Reanalysis of the Steubenville Data," *Inhalation Toxicology,* vol. 7 (January–February 1995), pp. 35–44.

16. Joel Schwartz, "PM sub 10, Ozone, and Hospital Admissions for the Elderly in Minneapolis-St. Paul, Minnesota," *Archives of Environmental Health,* vol. 49 (September 1994), p. 366.

17. Suresh Moolgavkar, E. Georg Luebeck, and Elizabeth L. Anderson, "Air Pollution and Hospital Admissions for Respiratory Causes in Minneapolis-St. Paul and Birmingham," *Epidemiology,* 1997, in press.

18. Joel Schwartz and Douglas W. Dockery, "Increased Mortality in Philadelphia Associated with Daily Air Pollution Concentrations," *American Review of Respiratory Diseases,* vol. 145 (March 1992), pp. 600–604.

19. Yuanzhang Li and H. Daniel Roth, "Daily Mortality Analysis by Using Different Regression Models in Philadelphia County, 1973–1990," *Inhalation Toxicology,* vol. 7 (January–February 1995), pp. 45–58.

20. Suresh H. Moolgavkar et al., "Air Pollution and Daily Mor-

tality in Philadelphia: The Philadelphia Story," *Epidemiology,* vol. 6 (September 1995), pp. 476–84.

21. Jonathan Samet, "Particulate Air Pollution and Mortality: The Philadelphia Story," *Epidemiology,* vol. 6 (September 1995), pp. 471–73.

22. Carol Browner, EPA administrator, prepared testimony of May 15, 1997.

23. Fred Rueter, personal telephone communication with the author, February 6, 1997.

24. Suresh Moolgavkar, M.D., of Fred Hutchinson Cancer Research Institute, Seattle, Washington, personal telephone communication with the author, January 15, 1997.

25. Ibid.

26. Ibid, March 28, 1997.

CHAPTER 4: A TALE OF SIX CITIES

1. Douglas W. Dockery et al., "An Association between Air Pollution and Mortality in Six U.S. Cities," *New England Journal of Medicine,* vol. 329 (December 9, 1993), pp. 1753–59.

2. Ibid, especially p. 1758.

3. Ibid.

4. Roger McClellan, president of Chemical Industry Institute of Toxicology, in Research Triangle Park, North Carolina, personal telephone communication with the author, April 11, 1997.

5. The median family income in Steubenville is about $7,000 lower than in Portage. See John Merline, "How Deadly Is Air Pollution?" *Consumers' Research,* vol. 80 (February 1997), p. 12.

6. Ibid.

7. See "Health Status Linked to Income, Lifestyle, and Type of Insurance," *Health Care Strategic Management,* vol. 14 (September 1996), p. 8.

8. Frederick W. Lipfert, "Estimating Air Pollution Mortality Risk from Cross-Sectional Studies: Prospective vs. Ecological Study Designs," *Particulate Matter: Health and Regulatory Issues,* American Waste Management Association Publication VIP-49, Pittsburgh, 1995, pp. 78–102.

9. Frederick W. Lipfert, of Brookhaven National Laboratory in Upton, New York, personal telephone communication with the author, May 13, 1997.

10. Frederick W. Lipfert and Ronald Wyzga, "Air Pollution

and Mortality: Issues and Uncertainties," *Journal of the Air & Waste Management Association*, vol. 45 (December 1995), pp. 949–66.

11. Ronald Wyzga, senior manager of Air Quality Risk and Health Assessment at Electric Power Research Institute, personal telephone communication with the author, January 13, 1997.

12. Fred Rueter, vice president, Consad, personal telephone communication with the author, February 6, 1997.

13. Kirk R. Smith, "Fuel Combustion, Air Pollution Exposure, and Health," *Annual Review of Energy and the Environment*, vol. 18 (1993), pp. 529–66.

14. Ibid.

15. Lipfert, personal telephone communication with the author, January 21, 1997.

16. Rueter, personal telephone communication with the author, February 6, 1997.

17. Robert Phalen, director of Air Pollution Health Effects Laboratory at University of California, Irvine, personal telephone communication with the author, January 13, 1997.

18. George T. Wolff and Roger O. McClellan, CASAC, letter to Carol M. Browner, May 16, 1994.

19. See John Merline, "EPA's Case of the Missing Data," *Investor's Business Daily*, February 21, 1997, p. A1. (Mentions one group.)

20. "Harvard Selects HEI to Review Six Cities PM Study," *Air Daily*, vol. 4 (April 14, 1997), p. 1.

21. Dan Greenbaum, president of Health Effects Institute, personal telephone communication with the author, March 5, 1997.

22. Mary D. Nichols, U.S. EPA, letter to Rep. Tom Bliley, undated.

23. Carol Browner, EPA administrator, prepared testimony before the U.S. House of Representatives Subcommittee on Health and Environment and Subcommittee on Oversight and Investigations, Committee on Commerce, May 15, 1997.

24. Laura Johannes, "Pollution Study Sparks Debate over Secret Data," *Wall Street Journal*, vol. 7, April 1997, p. B7.

25. Stuart Brody, *Sex at Risk* (New Brunswick, N.J.: Transaction, 1997), pp. 165–66, paraphrasing P. Skrabanek and J. McCormick, *Follies and Fallacies in Medicine* (Buffalo, N.Y.: Prometheus Books, 1990).

26. Neal Roth, of Roth Associates, Rockville, Md., personal telephone communication with the author, March 3, 1997.

27. Joel Schwartz, prepared statement before the Commit-

tee on Environment and Public Works, Subcommittee on Clean Air, Wetlands, Private Property and Nuclear Safety, February 5, 1997.

28. EPA, *Review of the National Ambient Air Quality Standards for Particulate Matter*, p. V-44.

29. Suresh Moolgavkar, M.D., of Fred Hutchinson Cancer Research Institute, Seattle, Washington, personal telephone communication with the author, March 28, 1997.

30. Antonio Ponce de Leon et al., "Effects of Air Pollution on Daily Hospital Admissions for Respiratory Disease in London Between 1987–88 and 1991–92," *Journal of Epidemiology & Community Health,* vol. 50 (supp. 1, April 1996), pp. S63–S70.

31. W. Dab et al., "Short Term Respiratory Health Effects of Ambient Air Pollutions: Results of the APHEA Project in Paris," *Journal of Epidemiology & Community Health,* vol. 50 (supp. 1, April 1996), pp. S42–S46.

32. Jordi Sunyer et al., "Air Pollution and Mortality in Barcelona," *Journal of Epidemiology & Community Health,* vol. 50 (supp. 1, April 1996), pp. S76–S80.

33. G. Touloumi, E. Samoli, and K. Katsouyanni, "Daily Mortality and 'Winter Type' Air Pollution in Athens, Greece—A Time Series Analysis within the APHEA Project," *Journal of Epidemiology & Community Health,* vol. 50 (supp. 1, April 1996), pp. S47–S51.

34. Bogdan Wojtyniak and Tomasz Piekarski, "Short Term Effect of Air Pollution on Mortality in Polish Urban Populations— What Is Different?" *Journal of Epidemiology & Community Health,* vol. 50 (supp. 1, April 1996), pp. S36–S41.

35. Antii Pönkä and Mikko Virtanen, "Asthma and Ambient Air Pollution in Helsinki," *Journal of Epidemiology & Community Health,* vol. 50 (supp. 1, April 1996), pp. S59–S62.

36. Claudia Spix and H. Erich Wichmann, "Daily Mortality and Air Pollutants: Finding from Köln, Germany," *Journal of Epidemiology & Community Health,* vol. 50 (supp. 1, April 1996), pp. S52–S58.

37. J.P. Schouten, J.M. Vonk, and A. de Graaf, "Short Term Effects of Air Pollution on Emergency Hospital Admissions for Respiratory Disease: Results of the APHEA Project on Two Major Cities in the Netherlands, 1977–89," *Journal of Epidemiology & Community Health,* vol. 50 (supp. 1, April 1996), pp. S22–S29.

38. L. Bachárová et al., "The Association between Air Pollution and the Daily Number of Deaths: Findings from the Slovak

Republic Contribution to the APHEA Project," *Journal of Epidemiology & Community Health,* vol. 50 (supp. 1, April 1996), pp. S19–S21.

39. Wojtyniak and Piekarski, "Short Term Effect," pp. S36–S41.

40. M.A. Vigotti et al., "Short Term Effects of Urban Air Pollution on Respiratory Health in Milan, Italy, 1980–89," *Journal of Epidemiology & Community Health,* vol. 50 (supp. 1, April 1996), pp. S71–S75.

41. Roth, personal telephone communication with the author, May 12, 1997.

CHAPTER 5: THE FINE PARTICLE FOLLIES

1. Fred Rueter, vice president, Consad, personal telephone communication with the author, February 6, 1997.

2. Carol Browner, EPA administrator, press conference of November 27, 1996.

3. As quoted in Philip Shabecoff, "Changes Proposed in Clean Air Law," *New York Times,* March 1984, p. 6.

4. American Lung Association, "Comments of the American Lung Association to the U.S. Environmental Protection Agency Regarding Proposed Decision: National Ambient Air Quality Standards for Particulate Matter" (61FR65638), March 12, 1997, pp. 1–2.

5. Public statement of Joel Schwartz, associate professor of environmental epidemiology, Harvard School of Public Health, and associate professor of medicine, Harvard Medical School, May 8, 1996.

6. U.S. EPA, *Air Quality Criteria for Particulate Matter,* EPA600/P-95/001a-cf (EPA, Office of Research and Development, National Center for Environmental Assessment: Research Triangle Park, N.C., 1996).

7. John D. Spengler et al., "Health Effects of Acid Aerosols on North American Children: Air Pollution Exposures," *Environmental Health Perspectives,* vol. 104 (May 1996), pp. 492–99.

8. Ibid., pp. 492–99.

9. C. Arden Pope et al., "Particulate Air Pollution as a Predictor of Mortality in a Prospective Study of U.S. Adults," *American Journal of Respiratory and Critical Care Medicine,* vol. 151 (March 1995), pp. 669–74.

10. For example, see "Particulates Get in Your Eyes," editorial, *Washington Times,* March 17, 1996, p. A22.

11. Suresh Moolgavkar, M.D., personal telephone commu-

nication with the author, January 15, 1997.

12. The study itself says 1979–1983, but they state they relied on the work of Fred Lipfert, who claims it was only from 1979–1981. See Frederick W. Lipfert, "Estimating Air Pollution Mortality Risks," from Cross-Sectional Studies: Prospective vs. Ecological Study Designs," *Particulate Matter: Health and Regulatory Issues*, American Waste Management Association Publication VIP-49, Pittsburgh, Pa., 1995, pp. 78–102.

13. Joseph Lyon, M.D., personal telephone communication with the author, January 22, 1997.

14. Lipfert, "Estimating Air Pollution Mortality Risks," pp. 78–102.

15. C. Arden Pope et al., "Particulate Air Pollution," p. 671, table 3.

16. U.S. EPA, "Statement by Mary Nichols, EPA Assistant Administrator, Office of Air and Radiation," April 2, 1997; see also, Joby Warrick, "EPA Concedes Error in Air Pollution Claim," *Washington Post*, April 3, 1997, p. A19.

17. Joel Schwartz, Douglas W. Dockery, and Lucas M. Neas, "Is Daily Mortality Associated Specifically with Fine Particles?" *Journal of the Air & Waste Management Association*, vol. 46 (October 1996), pp. 927–39, especially table 5.

18. American Lung Association, *Gambling with Public Health II*, p. 5.

19. Lyon, personal telephone communication with the author, February 12, 1997.

20. U.S. EPA, *Review of the National Ambient Air Quality Standards for Particulate Matter*, p. IV-9.

21. Lyon, personal telephone communication with the author, February 12, 1997.

22. U.S. EPA, *Air Quality Criteria for Particulate Matter*, p. 13-42, table 13-5.

23. Lipfert, personal telephone communication with the author, May 6, 1997.

24. Jack H. Shreffler, Richard L. Smith, and Jerry M. Davis, "An Initial Search for Association between Particulate Concentrations and Daily Mortality in Phoenix," presented at the Air and Waste Management Association Conference on the Measurement of Toxic and Related Air Pollutants, Research Triangle Park, North Carolina, April 29–May 1, 1997.

25. Carol Browner, EPA administrator, prepared testimony before the U.S. Senate Environment and Public Works Committee of February 12, 1997.

26. Browner, prepared testimony before the U.S. House of Representatives Subcommittee on Health and Environment and Subcommittee on Oversight and Investigations, Committee on Commerce, of May 15, 1997.

27. Daniel Menzel, personal telephone communication with the author, April 11, 1997.

28. See Michael Fumento, *Science under Siege* (New York: Morrow, 1993) pp. 45–77.

29. U.S. EPA, *Review of the National Ambient Air Quality Standards for Particulate Matter,* pp. V-69–73.

30. Joel Schwartz, prepared statement of February 5, 1997.

31. Joel Schwartz, prepared testimony before the U.S. Senate Environment and Public Works Committee, Subcommittee on Clean Air, Wetlands, Private Property, and Nuclear Safety, February 5, 1997.

32. John J. Godleski et al., "Death from Inhalation of Concentrated Ambient Air Particles in an Animal Model of Pulmonary Disease," *Proceedings of Second Colloquium on Particulate Air Pollution and Health,* Park City, Utah, in press, manuscript p. 7.

33. Douglas W. Dockery et al., "An Association between Air Pollution and Mortality in Six U.S. Cities," *New England Journal of Medicine,* vol. 324 (December 9, 1993), pp. 1753–59, table 1.

34. Godleski et al., "Death from Inhalation," p. 5.

35. Daniel Menzel, personal telephone communication with the author, April 11, 1997.

36. Daniel Menzel, prepared testimony before the Committee on Environment and Public Works, Subcommittee on Clean Air, Wetlands, Private Property, and Nuclear Safety, February 5, 1997.

37. Rueter, personal telephone communication with the author, February 6, 1997.

38. As quoted in Robert W. Crandall, Frederick H. Rueter, and Wilbur A. Steger, "Clearing the Air: EPA's Self-Assessment of Clean-Air Policy," *Regulation,* vol. 20 (1996), p. 43.

39. Deborah Sheiman Shprentz et al., *Breath-Taking: Premature Mortality due to Particulate Air Pollution in 239 American Cities* (New York: Natural Resources Defense Council, May 1996), pp. 32–34.

40. U.S. EPA, *Review of the National Ambient Air Quality Standards for Particulate Matter: Policy Assessment of Scientific and Technical Information OAQPS Staff Paper* (EPA-452/R-96-007, July 1996), p. VII-42.

41. American Lung Association, *Gambling with Public Health II*, p. 5.

42. Sverre Vedal, "Ambient Particles and Health: Lines That Divide," *Journal of Air & Waste Management*, vol. 47 (May 1997), pp. 551–81.

43. As quoted in Gary Polakovic, "Something Is Killing Americans," *Press-Enterprise* (Riverside, Calif.), July 14, 1996, p. A7.

44. See "EPA's PM2.5, Ozone Proposal Faces Rocky Road, But Many See Agency Prevailing," *Utility Environment Report*, December 6, 1996, p. 1.

45. Barry S. Levy, M.D., representing the American Public Health Association, prepared testimony before the U.S. House Subcommittee on Health and Environment and the Subcommittee on Oversight and Investigations, Committee on Commerce, May 8, 1997.

46. It was first used in 1888, according to *Webster's Ninth New Collegiate Dictionary* (Springfield, Mass.: Merriam-Webster Inc., 1990), p. 256.

47. Roger McClellan, personal telephone communication with the author, January 25, 1997.

48. Robert Phalen, personal telephone communication with the author, January 16, 1997.

Chapter 6: The Asthma Anomaly

1. Jeff Barker, "TV Ad Heralds Clean-Air Battle," *Arizona Republic*, February 12, 1997, p. A1.

2. As quoted in "Browner Says Will Not Be Swayed from Air Proposal," Reuters, February 21, 1997.

3. David Hawkins and Deborah Shprentz, "Peering through the Fog over Clean Air," *Washington Times*, December 8, 1996, p. B3.

4. For example, see statement of Judy Tucker, ALA News Conference on EPA Proposal on Particulate Matter Air Pollution, January 13, 1997.

5. "The Scary Spread of Asthma," cover of *Newsweek*, May 26, 1997.

6. As quoted in transcript of White House briefing, June 25, 1997.

7. Bob Herbert, "Bad Air Day," *New York Times*, February 10, 1997, p. A15.

8. As quoted in transcript of White House briefing, June 25, 1997.

9. Sharon Begley, Anne Underwood, and Daniel Glick, "Why Ebonie Can't Breathe," *Newsweek*, May 26, 1997, p. 61.

10. William O.C.M. Cookson and Miriam F. Moffatt, "Asthma: An Epidemic in the Absence of Infection?" *Science*, vol. 275 (January 3, 1997), p. 41.

11. Centers for Disease Control and Prevention, "Asthma Mortality and Hospitalization among Children and Young Adults—United States, 1980–83," *Morbidity and Mortality Weekly Report*, vol. 45, May 3, 1996, pp. 350–53, citing K.B. Weiss, P.J. Gergen, and D. K. Wagener, "Breathing Better or Wheezing Worse? The Changing Epidemiology of Asthma Morbidity and Mortality," *Annual Review of Public Health*, vol. 14 (1993), pp. 491–513.

12. Douglas W. Dockery et al., "Health Effects of Acid Aerosols on North American Children: Respiratory Symptoms," *Environmental Health Perspectives*, vol. 104 (May 1996), pp. 500–505.

13. CDC, "Asthma Mortality," pp. 350–53, figure 1.

14. Ibid., pp. 350–53.

15. David L. Rosenstreich et al., "The Role of Cockroach Allergy and Exposure to Cockroach Allergen in Causing Morbidity among Inner-City Children with Asthma," *New England Journal of Medicine*, vol. 336 (May 8, 1997), pp. 1356–63; see also Daniel Q. Haney, "Why Is Asthma So Bad in Cities? Cockroaches!" Associated Press, May 8, 1997.

16. Robert Phalen, director of Air Pollution Health Effects at University of California, Irvine, personal interview with the author, May 14, 1997.

17. "Physicians Applaud Clinton 'Courage' on Tough Air Standards," U.S. Newswire, June 25, 1997.

18. "Study Suggests Link between Pollution, Sudden Infant Death," Associated Press, July 10, 1997.

19. Brent Israelsen, "Study: Pollution Kills Many Infants," *Salt Lake Tribune,* June 12, 1997, p. 1.

20. Erin Kelly, "Air Pollution Puts Babies Susceptible to SIDS at Risk," Gannett News Service, July 10, 1997.

21. "SIDS, Air Pollution Linked," United Press International, July 10, 1997.

22. Tracey J. Woodruff, et al., " The Relationship between Selected Causes of Postneonatal Infant Mortality and Particulate Air Pollution in the United States," *Environmental Health Perspectives,* vol. 105 (June 1997), pp. 608–12.

23. "Short Order Science?" *Vital Stats,* Newsletter of the Statistical Assessment Service (July 1997), p. 3; David Murray, personal telephone communication with the author, based on his conversation with Julia Simmons, spokesperson for *Environmental Health Perspectives.*

24. As quoted in Sue MacDonald, *Cincinnati Enquirer,* "Body and Mind: Clean Air for Kids," July 2, 1997, p. 3.

25. Fred Rueter, personal telephone communication with the author, July 15, 1997.

CHAPTER 7: CASAC CAPERS

1. Carol Browner, EPA administrator, testimony before the U.S. Senate Environment and Public Works Committee, February 12, 1997.

2. Joby Warrick, "Clean Air Standards Opponents Circle the Backyard Barbecues," *Washington Post,* January 24, 1997, p. A1.

3. For example, see Browner's testimony of May 15, 1997, before the House of Representatives Subcommittee on Health and Environment and Subcommittee on Oversight and Investigations, Committee on Commerce.

4. Roger McClellan, president of Chemical Industry Institute of Toxicology, in Research Triangle Park, North Carolina, personal telephone communication with the author, April 11, 1997.

5. As quoted in Joel Bucher, "Weird Science: Did CASAC Really Support PM2.5?" *Issue Analysis of Citizens for a Sound Economy,* vol. 48 (April 9, 1997), p. 7, citing transcripts from CASAC meetings in May 1996, as prepared by County Court Reporters, Inc.

6. George T. Wolff, "The Particulate Matter NAAQS Review," *Journal of the Air and Waste Management Association,* vol. 46 (October 10, 1996), p. 926.

7. Clean Air Act Scientific Advisory Committee closure letter to EPA Administrator Carol M. Browner on the Staff Paper for Particulate Matter, June 13, 1996.

CHAPTER 8: HOLES IN THE OZONE CLAIMS

1. See http://ttnwww.rtpnc.epa.gov/naaqspro/o3list.htm, tables 1 and 2.

2. See Alan J. Krupnick and Deirdre Farrell, *Six Steps to a Healthier Ambient Ozone Policy,* Discussion Paper 96-13 (Washing-

ton, D.C.: Resources for the Future, 1996), p. 36, appendix B.

3. See Jacob Sullum, "What the Doctor Orders," *Reason,* vol. 27 (January 1996), p. 21.

4. Krupnick and Farrell, *Six Steps,* p. 36, appendix B.

5. Ibid., pp. 5–6.

6. As noted in Tom Spears, "Trees Contribute to Smog," *Ottawa Citizen,* February 14, 1994, p. A1. The report refers to "hydrocarbons," which includes all VOCs but some other gases as well.

7. National Research Council, *Rethinking the Ozone Problem in Urban and Regional Air Pollution* (Washington, D.C.: National Academy Press, 1991), p. 303.

8. Ibid., p. 244.

9. U.S. EPA, *Review of National Ambient Air Quality Standards for Ozone, Assessment of Scientific and Technical Information OAQPS Staff Paper,* EPA-452/R-96-007, Office of Air Quality Planning and Standards, Research Triangle Park, N.C., p. 61.

10. Ibid., p. 60.

11. Daniel Menzel, University of California, Irvine, personal telephone communication with the author, April 11, 1997.

12. Ari Patrinos, associate director, Health and Environmental Research, U.S. Department of Energy, letter to John Bachman, U.S. EPA, OAQPS, Research Triangle Park, N.C., January 4, 1995.

13. Randall Lutter and Christopher Wolz, "UV-B Screening by Tropospheric Ozone: Implications for the National Air Quality Standard," *Environmental Science and Technology/News,* vol. 31 (March 1997), pp. 142A–46A.

14. Donald H. Horstman et al., "Ozone Concentration and Pulmonary Response Relationships for 6.6-Hour Exposures with Five Hours of Moderate Exercise to 0.9, 0.10, and 0.12 PPM," *American Review of Respiratory Disease,* vol. 142 (November 1990), pp. 1158–63.

15. M. J. Hazucha, "Relationship between Ozone Exposure and Pulmonary Function Changes," *Journal of Applied Physiology* , vol. 62 (April 1987), pp. 1671–80.

16. U.S. EPA, *Review of the National Ambient Air Quality Standards for Ozone and Other Photochemical Oxidants* (Washington, D.C.: U.S. EPA, Office of Air Quality Planning and Standards, November 1987), p. VII-44.

17. Kenneth W. Chilton and Stephen Huebner, *Has the Battle against Urban Smog Become 'Mission Impossible'?* (St. Louis: Center

for the Study of American Business, November 1996), p. 9.

18. See Barbara B. Beck, Gradient Corporation, testimony before the Joint Hearing of the U.S. House of Representatives Health and Environment Subcommittee and the Oversight and Investigations Subcommittee, Commerce Committee, May 8, 1997.

19. George T. Wolff, "The Scientific Basis for a New Ozone Standard," *Environmental Manager,* vol. 2 (September 1996), pp. 27–32.

20. Ibid., p. 30.

21. Ibid., p. 32.

22. EPA, *Review of National Ambient Air Quality Standards for Ozone*, p. 158.

23. Ibid., p. 127.

24. Ibid., p. 130, table V-20.

25. Joby Warrick, "Panel Seeks Cease-Fire on Air Quality but Gets a War," *Washington Post,* February 6, 1997, p. A21.

26. James G. Donahue et al., "Inhaled Steroids and the Risk of Hospitalization for Asthma," *Journal of the American Medical Association,* vol. 277 (March 19, 1997), pp. 887–91. See also "Steroids Are Effective against Asthma, Study Says," *New York Times,* March 19, 1997, p. A12.

27. As quoted in Begley et al., "Yearning to Breathe Free," p. 63; see also Henry Milgrom et al., "Noncompliance and Treatment Failure in Children with Asthma," *Journal of Allergy and Clinical Immunology,* vol. 98 (December 1996), pp. 1051–57.

28. Kim Garr Ferguson, National Institutes of Health Program Budget Branch, Office of the Director, personal telephone communication with the author, May 2, 1997.

29. Morton Lippmann, M.D., testimony before the U.S. Senate Environment and Public Works Committee, February 5, 1997.

30. See Marlise Simons, "European Green Police Have Carrot but No Stick," *New York Times,* September 8, 1996, p. 3.

31. Roger McClellan, personal telephone communication with the author, May 6, 1997.

32. This can be found at http://134.67.104.12/G-DRIVE/O3PMRH/NRDIRS/PIECHRT.PDF.

33. See Evan Katz, "Tougher EPA Air Standards Could Hurt Farmers," States News Service, April 23, 1996.

34. Bob L. Vice, president, California Farm Bureau, Ameri-

can Farm Bureau Federation, testimony before the U.S. Senate Environment and Public Works Committee and Clean Air, Wetlands, Private Property and Nuclear Safety Subcommittee, April 29, 1997.

35. As quoted in George Anthan, "Critics: EPA Rules Dictate How to Plow," *Des Moines Register*, May 4, 1997, p. 1.

CHAPTER 9: THE IMPOSSIBLE DREAM

1. Clean Air Scientific Advisory Committee closure letter to EPA Administrator Carol Browner on the primary standard portion of the Office of Air Quality Planning and Standards (OAQPS) Staff Paper for Ozone, November 30, 1995.

2. Kenneth W. Chilton and Stephen Huebner, *Has the Battle against Urban Smog Become 'Mission Impossible'?* (St. Louis: Center for the Study of American Business, November 1996), p. 24.

3. 58 FR 51735, Presidential documents, *President of the United States,* Executive Order 12866 of September 30, 1993, Title 3- The President Regulatory Planning and Review, Part VIII, October 4, 1993, *Federal Register,* vol. 58, no. 190.

4. U.S. EPA, *Regulatory Impact Analysis for Proposed Particulate Matter: National Ambient Air Quality Standard* (Research Triangle Park, N.C.: December 1996), pt. 9, p. 145.

5. Ibid., pt. 9, p. 27, table 9.6.

6. Michael Hazilla and Raymond J. Kopp, "Social Cost of Environmental Quality Regulations: A General Equilibrium Analysis," *Journal of Political Economy,* vol. 98 (August 1990), pp. 853-73.

7. Robert W. Crandall, Frederick H. Rueter, and Wilbur A. Steger, "Clearing the Air: EPA's Self-Assessment of Clean-Air Policy," *Regulation,* vol. 20 (1996), p. 46.

8. U.S. EPA, *Regulatory Impact Analysis for Proposed Particulate Matter,* pt. 9, p. 27, table 9.6.

9. U.S. Department of Transportation, *Information Update of Value of Life and Injuries for Use in Preparing Economic Evaluations* (Washington, D.C.: DOT, March 1996), single sheet memorandum.

10. Alan Krupnick, Resources for the Future, testimony before the U.S. Senate Committee on Environment and Public Works, Subcommittee on Clean Air, Wetlands, Private Property, and Nuclear Safety, April 24, 1997.

11. EPA, "Review of the National Ambient Air Quality Standards for Particulate Matter," p. V-18.

12. Jonathan Samet, "Particulate Air Pollution and Mortality: The Philadelphia Story," *Epidemiology*, vol. 6 (September 1995), pp. 471–73.

13. See NRDC FAQs (frequently asked questions) at http://www.nrdc.org/faqs/aibrefaq.html#who.

14. Carol Browner, EPA administrator, testimony of May 15, 1997, before the U.S. House of Representatives Subcommittee on Health and Environment and Subcommittee on Oversight and Investigations, Committee on Commerce.

15. See Deborah Sheiman Shprentz et al., *Breath-Taking: Premature Mortality due to Particulate Air Pollution in 239 American Cities* (New York: Natural Resources Defense Council, May 1996), p. 73.

16. Marla Cone, "Grit in L.A. Air Blamed for 6,000 Deaths Yearly," *Los Angeles Times* (Washington ed.), May 9, 1996, p. A1.

17. Bureau of the Census, *Statistical Abstract: 1995*, p. 97, table 130.

18. Magnus Johannesson and Per-Olov Johansson, "To Be, or Not to Be, That is the Question: An Empirical Study of the WTP for an Increased Life Expectancy at an Advanced Age," *Journal of Risk and Uncertainty,* vol. 13 (September 1996), p. 171.

19. Using 1994 data, the Swedes had thirteen deaths per 100,000 people versus thirty-two for the United States. See graphic, "Deadly Road," *Washington Post,* May 24, 1997, p. A25.

CHAPTER 10: THE COST-BENEFIT CURVE

1. EPA, *Regulatory Impact Analysis for Proposed Particulate Matter: National Ambient Air Quality Standard* (Research Triangle Park, N.C., December 1996), pt. 7, p. 13.

2. Pechan and Associates, letter to EPA in EPA Docket A-95-54, November 22, 1996.

3. Thomas D. Hopkins, *Can New Air Standards for Fine Particles Live Up to EPA Hopes?* (St. Louis: Center for the Study of American Business, April 1997), p. 16.

4. Anne E. Smith et al., *Costs, Economic Impacts, and Benefits of EPA's Ozone and Particulate Standards* (Reason Public Policy Institute, June 1997), p. 25, table 1–6, and p. 26.

5. EPA, *Regulatory Impact Analysis for Proposed Particulate Matter,* p. 9-24, table 9.5.

6. Ray Squitieri, Office of Economic Policy, Department of the Treasury, memorandum to Mike Fitzpatrick, OIRA, December 19, 1996, p. 2.

7. Alicia Munnell, Council of Economic Advisers, draft memorandum to Art Fraas, December 13, 1996, p. 2.

8. Ibid.

9. Ibid., p. 4, table IV.

10. Bureau of the Census, *Statistical Abstract, 1996*, p. 112, table 155.

11. Council of Economic Advisers, "Questions about EPA Estimates," draft of December 6, 1996, p. 1.

12. Smith et al., *Costs, Economic Impacts, and Benefits*, p. 16.

13. Susan E. Dudley, *Comments on the U.S. Environmental Protection Agency's Proposed National Ambient Air Quality Standard for Ozone* (Fairfax, Va.: George Mason University, Center for the Study of Public Choice, March 12, 1997), p. II-8, and appendix A, p. 42; see also Susan E. Dudley, vice president and director of environmental analysis at Economists Incorporated of Washington, D.C., testimony before the U.S. Senate Committee on Environment and Public Works Subcommittee on Clean Air, Wetlands, Private Property, and Nuclear Safety, April 24, 1997.

Numerous other studies have been done, also with sobering results. In a 1989 report for the Office of Technology Assessment (OTA), Alan Krupnick and Raymond Kopp estimated the cost of reducing just VOCs for all nonattainment areas by just 35 percent. Even this level of abatement, however, would not solve the ozone problem in these areas; rather the authors projected it would bring only one-third of the areas that were in mild violation of the ozone standard in 1985 into attainment by 2004. Yet just for this, they found the cost (converted to 1995 dollars) to be $7.8 billion to $11.9 billion a year (Raymond J. Kopp and Alan J. Krupnick, "Controlling Emissions of Volatile Organic Compounds," in U.S. Congress, Office of Technology Assessment, *Catching Our Breath: Next Steps for Reducing Urban Ozone*, OTA-O-412 [Washington, D.C.: U.S. Government Printing Office, July 1989], p. 140).

Meanwhile, the Center for the Study of American Business did its own analysis in 1996, finding that reducing VOCs by 40 percent in nonattainment areas by 2005 would be approximately $11.2 billion a year in 1995 dollars, putting it in the same ballpark as the OTA estimate. Kenneth W. Chilton and Stephen Huebner, *Has the Battle Against Urban Smog Become 'Mission Impossible'?* (St. Louis: Center for the Study of American Business, November 1996), p. 13 and p. 14, table 2.

Clearly a move to bring even more areas into attainment would entail costs grossly disproportionate to the savings. But how great? Sierra Research of Sacramento, California (no relation to the Sierra Club) did a study on behalf of the American Petroleum Institute (API) focusing just on the area around Chicago, which is currently not in attainment. Sierra estimated that meeting an 0.09 ppm standard would cost an estimated $1.8 billion to $4.4 billion a year. Meeting the tighter 0.08 ppm standard would set Windy City denizens back an estimated $2.5 billion to $7 billion annually. Prasad Rao and Thomas J. Lareau, "The Monetary Benefits of an Eight-Hour 0.08 ppm Ozone Standard in Chicago, Research Study No. 085 (Washington, D.C., American Petroleum Institute, August 1996), p. xii.

As to the benefits, OTA pegged those at $2.4 billion a year. This would mean spending about $3 to $5 to save each dollar under the OTA calculation, and about $4 to save each dollar under the CSAB estimate (Chilton and Huebner, *Has the Battle against Urban Smog*, p. 14, table 2). Sierra in its study did no benefit analysis, but API used figures in the study to do so itself, for only the tighter standard. It found costs would range anywhere from 80 to 210 times the benefits (ibid., p. 17, table 3). Obviously, the results of one area cannot be extrapolated to the entire nation. For some areas the cost-benefit ratio will assuredly be less; for others more. But it is sobering to realize that depending on how broadly the Chicago area and the population therein are defined, this could cost each person more than $1,000 a year. Nobody has yet figured out what the ALA's proposed 0.07 ppm standard could cost, but suffice it to say a lot of people in the area around Chicago might just as well sign over their paycheck to the U.S. government.

14. Barry S. Levy, M.D., testimony of May 8,1997; see also, "Comments of the American Public Health Association," to Environmental Protection Agency, Air Docket (6102), Attn: Docket no. A-95-54, March 11, 1997.

15. Indur M. Goklany, "Richer Is Cleaner: Long-Term Trends in Global Air Quality," in Ronald Bailey, ed., *The True State of the*

Planet (New York: Free Press, 1995), p. 341.

16. W. Kip Viscusi, "The Dangers of Unbounded Commitments to Regulate Risk," in Robert Hahn, ed., *Risks, Costs, and Lives Saved: Getting Better Results from Regulation* (New York: Oxford University Press, 1996), p. 162.

17. Calculated from Ralph L. Keeney and Kenneth Green, *Estimating Fatalities Induced by Economic Impacts of EPA's Proposed Ozone and Particulate Standards* (Los Angeles: Reason Public Policy Institute, June 1997), p. 7, table A, and p. 8.

18. Thomas Lareau, American Petroleum Institute, personal telephone communication with the author, March 6, 1997.

19. Ibid.

20. John Schneibel, personal telephone communication with the author, March 7, 1997.

21. See Gregg Easterbrook, "They Stopped the Sky from Falling," *Washington Monthly,* vol. 27 (May 1995), p. 34.

22. As quoted in Agis Salpukas, "Green Power Wanes, but Not at the Grass Roots," *New York Times*, March 9, 1997, p. D1.

23. As quoted in "EPA's PM2.5, Ozone Proposal Faces Rocky Road, But Many See Agency Prevailing," *Utility Environment Report*, December 6, 1996, p. 1.

24. Carol Browner, EPA administrator, testimony before the Veterans Affairs, Housing and Urban Development, and Independent Agencies Subcommittee of the U.S. House Appropriations Committee, April 15, 1997.

25. Gordon Hester, environmental risk program, EPRI, personal telephone communication with the author, March 7, 1997.

CHAPTER 11: BARBECUE POLICE

1. Carol Browner, EPA administrator, testimony before the House of Representatives Subcommittee on Health and Environment and Subcommittee on Oversight and Investigations, Committee on Commerce, May 15, 1997.

2. As quoted in Eric Niiler, "EPA Chief: 'Scare Tactics' Mar Clean-Air Debate," *Patriot Ledger,* February 25, 1997, p. 6.

3. As quoted in Christine Magnotta, "Environmentalists Target Maine for Clean-Air Ads," *Bangor Daily News*, February 15, 1997.

4. As quoted in "EPA Plans to Curb Emissions of Lawn Mowers," *National Home Center News,* May 23, 1994, p. 12.

5. Sam Walker, "Beware: Your Lawnmower Is an Environmental Enemy," *Christian Science Monitor*, August 12, 1996, p. 3.

6. Department of Defense, "Comments on the Proposed Rules for National Ambient Air Quality Standards for Particulate Matter and Ozone and Interim Implementation Policy," March 12, 1997.

7. See "New U.S. Motorboat Rule to Slash Emissions," Reuters Financial Service, August 5, 1996.

8. David Foster, "Green Guilt or Gracious Living?" *Los Angeles Times*, December 11, 1994, p. B4.

9. Richard Wiles, "Are Tougher Air Standards Necessary?" *Washington Times*, February 2, 1997, p. B3.

10. Data from American Foundrymens' Society and the Barbecue Industry Association.

11. David Foster, "Green Guilt," p. B4.

12. Roger McClellan, president of Chemical Industry Institute of Toxicology, Research Triangle Park, N.C., personal telephone communication with the author, January 25, 1997.

13. Carol Browner, testimony of May 15, 1997.

14. Frank E. Kruesi, assistant secretary for transportation policy, memorandum to Sally Katzen, November 20, 1996.

15. Robert Phalen, director of Air Pollution Health Effects Laboratory, University of California, Irvine, personal telephone communication with the author, January 16, 1997.

16. Small Business Regulatory Enforcement Fairness Act, Pub. L. No. 104-121, sec. 241, 110 Stat. 864 (1996).

17. The allusion, for those who don't get it, is to the famous line in "The Treasure of the Sierra Madres," with Humphrey Bogart, in which the bandidos pretending to be police say, "Badges? We don't need no steenking badges." Actually, the line wasn't quite uttered that way but has become immortalized nonetheless, as has another unspoken line, "Play it again, Sam," from another Bogart movie.

18. Jere Glover, chief counsel for advocacy, Small Business Administration, memorandum to Carol Browner, November 11, 1996.

19. K. C. Tominovich, personal telephone communication with the author, January 31, 1997.

20. As quoted in "Air Pollution," *BNA's Environment Report*, vol. 27 (March 21, 1997), p. 2268.

21. Ibid.

22. Peter Homer, personal telephone communication with the author, April 17, 1997.

23. Jere Glover, memorandum to Carol Browner, November 11, 1996.

24. "Issues concerning the PM and Ozone Rules," USDA memorandum from the Office of the Secretary, undated.

25. Jere Glover, memorandum to Carol Browner, November 11, 1996.

26. Carol Browner, testimony of May 15, 1997.

CHAPTER 12: DISTRACTIONS AND DOUBLE STANDARDS

1. John J. Fialka, "Panel Judging EPA's Proposed Air Regulations Receives Most of Its Funding from the Regulated," *Wall Street Journal*, January 16, 1997, p. A20.

2. "Our Breath-Taking Air," *U.S. News & World Report*, May 20, 1996, p. 15; Marla Cone, "Grit in L.A. Air Blamed for 6,000 Deaths Yearly," *Los Angeles Times* (Washington Ed.), May 9, 1996, p. A1; and Gary Lee, "Air Pollution Tied to 64,000 Premature U.S. Deaths," *Washington Post*, May 9, 1996, p. A4.

3. Our Breath-Taking Air," p. 15; "Microscopic Killers," *New York Times*, May 12, 1996, p. E12; "Fighting a Microscopic Killer," *Atlanta Journal–Atlanta Constitution*, May 11, 1996, p. A10; Mark Jaffe, "Dirty Air's Death Toll Estimated at 64,000," *Philadelphia Inquirer*, May 9, 1996, p. A1.

4. For example, Douglas W. Dockery, Joel Schwartz, and John D. Spengler, "Air Pollution and Daily Mortality: Associations with Particulates and Acid Aerosols," *Environmental Research*, vol. 59 (December 1992), pp. 362–73.

5. John Merline, "EPA Boosters on the Gov't Tab," *Investor's Business Daily*, January 28, 1997, p. A1.

6. John B. Garrison, managing director, American Lung Association, and John H. Adams, executive director, Natural Resources Defense Council, "Responding to EPA Grants," *Investor's Business Daily*, February 7, 1997, p. A34.

7. See Peter Asmus and Linda Dailey Paulson, "A Concrete Concern," *San Francisco Examiner*, April 23, 1995, p. B1.

8. James T. Bennett and Thomas J. DiLorenzo, "Commercialization of America's Health Charities," *Society*, vol. 34 (May–June 1997), p. 70, citing the Harrisburg, Pennsylvania, *Patriot* newspaper.

9. Don Hopey, "Donors Annoyed about Support," *Pittsburgh Post-Gazette*, September 28, 1994, p. C1.

10. As quoted in Bennett and DiLorenzo, "Commercialization," p. 70.

11. Doug Bandow, "Cost-Benefits of Enviro-Purity," *Washington Times*, April 3, 1997, p. A13.

12. American Lung Association, *Gambling with Public Health II: Who Loses under New Health Standards for Particulate Matter* (Washington, D.C.: ALA, January 1997), p. 2.

13. See, generally, American Lung Association, "Comments of the American Lung Association to the U.S. Environmental Protection Agency."

14. Deborah Sheiman Shprentz et al., *Breath-Taking: Premature Mortality due to Particulate Air Pollution in 239 American Cities* (New York: Natural Resources Defense Council, May 1996), p. ii.

15. Robert Phalen, personal telephone communication with the author, January 16, 1997.

16. As stated in EPA, "EPA's Proposed Ozone and Particulate Matter Public Health Standards," memorandum, November 21, 1996, p. 2.

17. Joel Schwartz, "Stuff and Nonsense in Regard to a Ruling," letter, *Wall Street Journal*, March 19, 1997, p. A19.

CHAPTER 13: THE CLINTON ADMINISTRATION

1. H. Josef Hebert, "EPA Air Plan Set despite Concerns at Cabinet Level," *Denver Post*, March 15, 1997, p. 16A.

2. George T. Frampton, Jr., Department of the Interior, memorandum to Mary Nichols and Sally Katzen, November 26, 1996, p. 4.

3. Frank E. Kruesi, memorandum of November 20, 1996.

4. Jack Gibbons, Office of Science and Technology Policy, memorandum to Sally Katzen, November 21, 1996.

5. Rosina Bierbaum, Environment Division, Office of Science and Technology Policy, draft memorandum to Sally Katzen, November 15, 1996, p. 1.

6. John Beale, EPA, "Response to Chairman Bliley's Letter," memorandum to Art Fraas, undated, p. 2.

7. See testimony of Sally Katzen and Mary Nichols before the joint hearing of the Health and Environment and the Oversight and Investigation Subcommittees of the U.S. House Commerce Committee, April 17, 1997.

8. Ibid.

9. See H. Josef Hebert, "Congressman Says EPA Quashes Interagency Criticism," Associated Press, February 26, 1997.

10. Frederick W. Lipfert, personal telephone communication with the author, March 27, 1997.

11. Department of Agriculture, "National Ambient Air Quality Standards: Ozone and Particulate Matter," undated memorandum.

12. Department of Defense, "Comments on the Proposed Rules for National Ambient Air Quality," pp. 2, 5.

13. E-mail from Jean Vernet, Department of Energy, to Arlene Anderson et al., "03/PM Comments Not to Be Transmitted," March 11, 1997.

14. Notes from interagency NAAQS briefing, written on Executive Office of the President, OSTP letterhead, November 12, 1996, source unnamed.

15. Jack Gribbon, OMB press office, personal telephone communication with the author, May 12, 1997.

CHAPTER 14: FIRST DO NO HARM

1. Carol Browner, testimony before the U.S. House of Representatives Subcommittee on Health and Environment and Subcommittee on Oversight and Investigations, Committee on Commerce, May 15, 1997.

2. EPA, *Review of the National Ambient Air Quality Standards for Particulate Matter: Policy Assessment of Scientific and Technical Information OAQ PS Staff Paper* (EPA-452/R-96-007, July 1996), p. VII-43.

3. George D. Thurston et al., "Respiratory Hospital Admissions and Summertime Haze Air Pollution in Toronto, Ontario: Consideration of the Role of Acid Aerosols," *Environmental Research,* vol. 65 (May 1994), pp. 271–90.

4. C. Arden Pope, personal telephone communication with the author, May 2, 1997.

5. Daniel Defoe, *Defoe's History of the Great Plague in London: A Journal of the Plague Year* (Boston and London: Ginn & Co., 1895) p. 125.

6. K. C. Shaw, Geneva Steel Corporation, personal telephone communication with the author, January 21, 1997.

7. Roger McClellan, personal telephone communication with the author, April 11, 1997; see, also, Roger McClellan,

"Time to Move Beyond the Regulatory Lamp Post," *Particulars,* July 1996, pp. 2, 11.

8. Frederick W. Lipfert, personal telephone communication with the author, February 6, 1997.

9. Robert Phalen, personal telephone communication with the author, January 16, 1997.

10. Ibid.

11. EPA, *Brochure on National Air Quality: Status and Trends, EPA-454/F-96-008* (Federal Triangle Park, N.C.: EPA Office of Air Quality, October 1996), p. 8.

12. EPA, *Review of the National Ambient Air Quality Standards for Particulate Matter,* p. IV-11.

13. See Kurt Kleiner, "Clean Air Plans Steeped in Confusion," *New Scientist,* vol. 153 (February 15, 1997), p. 8.

14. Roger McClellan, personal telephone communication with the author, January 25, 1997.

15. Daniel Menzel, personal telephone communication with the author, April 11, 1997.

CHAPTER 15: SCIENCE AND REALITY

1. Carol Browner testimony before the Senate Environment and Public Works Committee, February 12, 1997.

2. Robert Hahn, resident scholar, American Enterprise Institute, personal telephone communication with the author, May 2, 1997.

3. Memorandum from Frank E. Kruesi, November 20, 1996.

4. Congressional Review Act of 1996, 5 U.S.C. 801(a)(3).

5. See H. Joseph Hebert, "Court Asked to Block Tougher Air Standards," Associated Press, July 18, 1997.

6. Unfunded Mandates Reform Act of 1995, 2 U.S.C.S. sec. 1501 (L. Ed. Supp. 1996).

7. See Andrew Ferguson, "Are the Democrats Going Nuts? An Inquiry," *Weekly Standard,* vol. 1 (September 18, 1995), p. 38.

8. Representative John Dingell, written communication to President William J. Clinton, April 24, 1997, p. 1.

9. Representative Rick Boucher, written communication to President William J. Clinton, April 22, 1997, pp.1, 2.

10. Times Mirror Survey as reported in *The American Enterprise,* vol. 6 (March–April 1995), p. 109.

Index

Abortion, 15
Acid rain, 70
ACS II, 19, 30–31, 33
Agriculture Department, 76, 84, 87–88
Air pollution or quality, *see* Pollution
American Farm Bureau, 57
American Lung Association, 37, 39, 90; and asthma prevention, 41; changing goals, 79–80; demands for PM2.5 standards, 5, 28–29, 80–82; lawsuits against EPA, 2, 78–79; and ozone standards, 49–50, 52
American Public Health Association, 49
Animal studies, 35–36
Association for Responsible Thermal Treatment, 80
Asthma, 39, and cockroaches, 42; effect of PM2.5 regulations on, 57; and ozone levels, 54–55; and pollution, 41;

race-related, 41; rates in Europe, 40–41; rising among children, 40
Atomic energy, 69
Automobile emission testing program, 4, 80

Barbecues, 72, 73, 74
Beale, John, 86
Begley, Sharon, 40
Benefits, *see* Cost-benefit analyses
Best practices document (OIRA), 85–86
Bierbaum, Rosina, 85
Billings, Paul, 80
Birmingham, Alabama, 13–14, 16, 26, 82
Bliley, Thomas, 24, 85–86
Boston, 32, 102
Boucher, Rick, 98–99
Breath-Taking (NRDC), 5, 36–37, 77, 82
Brody, Stuart, 25
Bronchitis, 64

About the Author

MICHAEL FUMENTO, a noted science correspondent and author, is a resident fellow at the American Enterprise Institute for Public Policy Research. He was the 1994 Warren T. Brookes Fellow in Environmental Journalism at the Competitive Enterprise Institute, a fellow with Consumer Alert, and a science writer for *Reason* magazine. His commentaries have appeared in the *New York Times*, the *Wall Street Journal*, the *Los Angeles Times*, the *Christian Science Monitor*, and other newspapers, and he has lectured on science and health issues throughout the world. He is also the author of *The Myth of Heterosexual AIDS* (1990, rev. ed. 1993), *Science under Siege* (1993), and *The Fat of the Land* (September 1997).

This book was edited by
Cheryl Weissman and the publications staff
of the American Enterprise Institute.
The index was prepared by Nancy Rosenberg.
The text was set in New Baskerville.
Jennifer Lesiak set the type,
and Edwards Brothers, Incorporated,
of Ann Arbor, Michigan,
printed and bound the book,
using permanent acid-free paper.

The AEI Press is the publisher for the American Enterprise Insti-
tute for Public Policy Research, 1150 Seventeenth Street, N.W.,
Washington, D.C. 20036; *Christopher DeMuth*, publisher; *Dana
Lane*, director; *Ann Petty*, editor; *Leigh Tripoli*, editor; *Cheryl
Weissman*, editor; *Jennifer Lesiak*, production manager.